Digital Techniques in Broadcasting Transmission

Digital Techniques in Broadcasting Transmission

Robin Blair

**Focal
Press**

Boston Oxford Auckland Johannesburg Melbourne New Delhi

Focal Press is an imprint of Butterworth–Heinemann.

Copyright © 1999 by Butterworth–Heinemann

 A member of the Reed Elsevier group

Library of Congress Cataloging-in-Publication Data
Blair, Robin, 1942–
    Digital techniques in broadcasting transmission / Robin Blair.
      p.   cm.
    Includes bibliographical references.
    ISBN 0-240-80366-3 (pbk. : alk. paper)
    1. Digital television.   2. Television--Transmitters and transmission.   I. Title.
    TK6678.B53   1999
    621.388--dc21   99-18132
                                            CIP

British Library Cataloguing-in-Publication Data
A catalogue record for this book is available from the British Library.

The publisher offers special discounts on bulk orders of this book.
For information, please contact:
Manager of Special Sales
Butterworth–Heinemann
225 Wildwood Avenue
Woburn, MA 01801-2041
Tel: 781-904-2500
Fax: 781-904-2620

For information on all Butterworth-Heinemann publications available, contact our World Wide Web home page at: http://www.bh.com

10 9 8 7 6 5 4 3 2 1

Printed in the United States of America

Contents

Preface

Digital techniques have been used in TV and sound studios for years, with quite spectacular success. Now the time has come to introduce digital techniques into the transmission chain. However, few people engaged in that side of broadcasting have had much exposure to the basic principles and techniques that underlie digital transmission systems. To fill this gap, this book is intended to be a primer on the theory of digital transmission with an emphasis on its application to broadcasting.

The book is structured around the approach of discussing the principles and then showing how they are being applied in practice in the systems now emerging. The definition of a transmission system is fairly broad. In this context, video and audio compression systems reduce the bit rates required to well-below studio standards and make their transmission practical within the bandwidths available for broadcasting. In the author's opinion they are undoubtedly part of the transmission system, and an extremely interesting one at that. Hence, they are treated in some detail, especially the famous MPEG video compression standards whose influence is spreading into myriad applications as well as broadcasting.

The practical systems described in the book encompass the ATSC 8VSB TV Standard developed in the United States, the European Digital Video Broadcasting (DVB) standards, Eureka 147 digital audio broadcasting, and some satellite applications. The treatment focuses on systems that are, or are likely to become, recognized world standards. There is a short history of these in Chapter 1.

Several of the chapters are aimed at the practical aspects of engineering digital systems in the field. The engineering environment is similar to, but not the same as, what we have experienced in analog transmission. Some of what was regarded as vitally important in analog transmission now hardly matters at all, but naturally the digital medium does introduce much that is new. It is the author's opinion, however, that once the principles are understood, station engineering is probably simpler with digital than it has been with analog.

It will be a good many years before the analog channels disappear completely, and in the meantime analog and digital will have to live together. Many existing operators will find themselves having to transmit their original analog service and the new digital services in parallel. Hence, the book explores what is required in these situations. It contains background material on sharing antennas and channel combining, and particularly adjacent channel combining, which may be very useful techniques during the transition period to full digital.

The book generally avoids the use of mathematics except at a very basic level. However, one's appreciation of some of the finer theoretical details is much enhanced by an understanding of the Fourier transform. A short appendix explains the use of this in a way that, it is hoped, will be readily understood by non-mathematical readers.

Chapter 1

Introduction

Why Digital?

There cannot be many human activities left on this earth where digital techniques have not taken over from older analog methods and vastly improved the end results. Thus we see applications from telephony to the control of automobile engines that have taken the digital route. Probably the one that is most obvious to the general public is the way the compact disk has almost totally replaced the vinyl record in a remarkably short time. As an engineer I can remember another – the way the electronic calculator displaced the slide rule that had been an icon of our profession for centuries.

Yet broadcasting systems have remained staunchly analog, both for radio and TV, almost certainly because very good compression algorithms and the means to implement them have only recently become available. Compression is the process by which the digital bit rate required to transmit high-quality audio or video can be reduced to some value compatible with the channel bandwidths available to broadcasting services. A compact disk player typically utilizes a bit rate of about 1.5 Mb/s, and in an uncompressed form video transmission would require over 200 Mb/s. (Mb/s is short for megabits per second. Readers unfamiliar with such terms are referred to the glossary at the end of the book.)

Attempts to reduce the bandwidth required for video transmission began not long after World War II, and some of the techniques described herein were known in the 1960s. However, their implementation had to await the other digital revolution in the development of large-scale integration chips with sufficient computation power to run the algorithms in real time. The advances have been dramatic. Most major countries have announced their intention to have digital TV and radio services operating by the year 2000. For this they will rely on recently developed standards. The standards have had a short, but intense, history of development. Let us look briefly at how the major ones have come into being.

Digital TV in the United States

During the 1980s, quality was becoming something of an issue for conventional broadcasters. Typical domestic video displays compared very poorly with what people were seeing in the cinema, and the emerging compact disks were seen as a real threat to music programs on radio. There were intense efforts to develop high definition TV (HDTV), but all required a relatively large bandwidth and, it seemed, could be delivered only by cable. The broadcasters naturally saw this as a threat to their businesses, and in 1987 a large group in the United States persuaded the FCC to set up the Advisory Committee on Advanced Television Systems (ACATS).

ACATS set itself the task of testing candidate HDTV systems and selecting a North American Standard by June 1990. All the systems initially submitted to it were analog, but one digital system entered the fray just before the target cut-off date. ACATS sought an extension and by March 1992 had four proposals for digital systems and only two for analog, one of which was subsequently withdrawn. After intensive testing ACATS came down firmly on the side of digital transmission and recommended that the four proponent groups be given time to refine their systems. Seven of the major organizations involved then elected to form a cooperative known as the Grand Alliance, with the aim of combining the best features of the four systems to create the best possible standard. At this time also there was an intensive worldwide effort to develop a universal video and audio compression system. The result became the famous MPEG standard (from the semi-official title of the committee, Moving Picture Experts Group). The Grand Alliance adopted much of the MPEG standard for incorporation into its system.

Two and a half years of intensive testing and refinement followed. The Advanced Television System Committee (ATSC) Standard was agreed in November 1995, and in December 1996 the FCC endorsed this as the digital TV (DTV) standard for the United States.

The DVB Project

There had been a number of attempts in Europe to devise an enhanced TV standard using analog techniques. One of the more famous was the MAC (multiple analog components) scheme. This compressed the components of the video signal in time and transmitted them in sequence, rather than in parallel, as happens in PAL or NTSC, and thus avoided many of the artifacts caused by mixing between the video components. It looked so much like the system of the future that at one time the European Commission legislated to make it compulsory in all new TV sets.

In 1990, however, General Instruments had demonstrated its Digicipher video compression system, which showed much promise for truly useful results in digital transmission. The concepts were adopted by MPEG, and that committee released its first standard in 1993. These developments made it obvious that the future lay with digital, and in 1993 the DVB Project was born.

DVB is an alliance of manufacturers, broadcasters, government agencies, and research institutions that has come to have over 300 members. Its aim is to develop digital TV broadcasting signal formats for delivery by terrestrial transmitters, cable systems, and satellites. There is a very high degree of commonality among all of these formats so that consumers may very easily receive any or all of them. The European Telecommunications Standards Institute (ETSI) has adopted the formats developed by DVB as European Standards.

4

The Eureka 147 Project

Like DVB, Eureka 147 is a collaborative project being undertaken by a consortium of organizations with the aim of developing a world standard for digital audio broadcasting (DAB). It has over 40 members in total, mostly drawn from Europe but with significant contributions being made from other countries, notably Canada from the early days and lately Japan. The consortium was first formed in 1987. It has developed a multichannel sound broadcasting system with the compression based on the work of MPEG. The system was adopted as an ETSI Standard in early 1995, and the first commercial services went to air in Stockholm and London at the end of that year.

IBOC and IBAC DAB

As a multichannel system, Eureka 147 does not particularly suit broadcasters in the United States, where most operators are licensed to transmit only one program. It is more suited to the situation where big public or private organizations provide integrated suites of programs, such as those that exist in many European countries and Canada. Accordingly, developers in the United States are working on systems known as in band on channel (IBOC) or in band adjacent channel (IBAC). These are specifically designed to permit an existing broadcaster to transmit a single digital service either within his existing channel or the one adjacent to it, which will almost always be vacant in his allocated service area. The great difficulty with this situation is to control interference between the new digital and existing analog services, and at the time of writing, none of several experimental systems has reached the stage of being adopted as a national standard.

A New Engineering Paradigm

Thomas Kuhn in his book *The Structure of Scientific Revolutions* first adopted the word *paradigm* to describe the set of theories, beliefs, and tests, adopted by the practitioners in some field of science from time to time. Analog broadcasting transmission has certainly had its own paradigm. For example, we have seen the importance placed on a myriad of waveform tests in TV testing and the great deal of attention given to impedance matching in antenna and transmission line systems in station design.

The digital era will require a new paradigm. We will show in this book that waveform testing is no longer appropriate except in one particular. The reader will see that the digital systems are specifically designed to resist echoes far worse than those that arise from small mismatches in antenna systems. At the same time, the transmission engineer will have to become familiar with some new requirements, for example, describing voltage and power in statistical terms rather than as defined by standard waveforms.

However, there is nothing magical about digital transmission, and anyone with some knowledge of broadcasting will be able to understand the principles easily. At worst, one may have to come to terms with the Fourier transform to switch one's thinking between the time and the frequency domains. Even this, however, is a concept with which many broadcasting engineers are familiar, particularly those who have worked in television.

This book lays out systematically the principles that lie behind the current standards for digital broadcasting, and will be most useful to those who work in or have an interest in the current analog transmission systems.

Chapter 2

Sending by Numbers: Basic Digital Transmission

This chapter describes the essential concepts of digital transmission. It is shown that the important concepts apply to even the simplest baseband systems. However, the evolution from simple systems to the seemingly very complex applications in broadcasting is quite direct. We will see that many of the parameters that appear in broadcasting systems standards have their genesis in simple baseband data transmission.

Elementary Baseband Transmission

Digital transmission is all about sending and receiving symbols. The symbols belong to a limited and defined set. Each symbol has a very specific value, usually a number or letter. It is the task of the receiver to decide which symbol has been sent and to assign the correct value to it. The analogy with human spoken communication is compelling. There, the speaker and listener share a vocabulary of a finite number of words, albeit possibly a very large number. The task of the listener is to decide which word has been spoken, amongst the likely possibilities. Thus, even at a noisy cocktail party, say, the listener may hear a word imperfectly but still be able to decide which word it is. If he or she generally makes correct decisions, then good understanding is possible even through the noisy communication channel that one inevitably finds at cocktail parties.

The simplest digital transmission one can imagine would be for transmitting binary digits. These can take only the values 1 and 0, and so only two different symbols are necessary, and a good choice might be a positive voltage pulse for the value 1 and a negative voltage pulse for 0. An elementary diagram for this system and the transmission waveform for an arbitrary stream of data is shown in Figure 2.1.

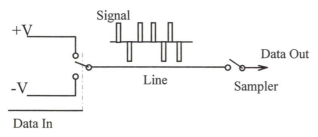

Figure 2.1: Elementary Digital Transmission System

Pulse Shaping and Bandwidth

Some things are immediately obvious. The more the pulses are narrowed and squeezed together in time, the greater the data transmission rate. The limitation here however is the bandwidth that can be afforded to the system. There is a well-known inverse relationship between the width of a voltage pulse and the bandwidth required for its transmission, and the pulse will be significantly distorted if that bandwidth is not available. In digital transmission, however, we can control the distortion of the pulses so that the receiver is able to distinguish one type from another—between positive and negative in this case. Second, the lack of bandwidth will cause some smearing of the pulse shape in time, causing leading and trailing pulses to overlap, and create "intersymbol interference." Again, this is of no consequence if it does not add any ambiguity to distinguishing between the two classes of symbol. Given those considerations, let us look at how our elementary system fares if the bandwidth is strictly limited.

The ideal limited bandwidth channel would have, at baseband, a frequency response flat up to some frequency F and zero from there on. If we were to apply a pulse to this, Fourier transform theory shows that the output voltage waveform at the receiver would take the form given by $\sin(kt)/t$, where $k = 2\pi F$. The frequency response and pulse waveform are shown in Figures 2.2.a and 2.2.b respectively. Although this resulting waveform is far from the ideal pulse the transmitter may have sent, it has a clearly defined peak and the receiver could distinguish positive from negative by sampling at the peak.

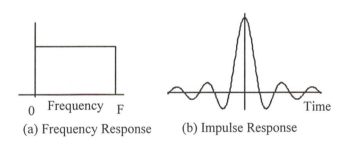

| (a) Frequency Response | (b) Impulse Response |

Figure 2.2: Channel Frequency and Impulse Response

Now consider two pulses to be transmitted in succession; to be specific a positive pulse followed by a negative one. The manner in which the two output symbols might overlap is shown in Figure 2.3. Notice that if we choose the time between the pulses correctly, the times at which each attains its peak value can be arranged to fall at a time at which the waveform of the other passes through zero. Hence, as the receiver takes samples at these times, the sampled peak value of each symbol is not influenced by the presence of the other. The intersymbol interference is thus zero at these sampling instances.

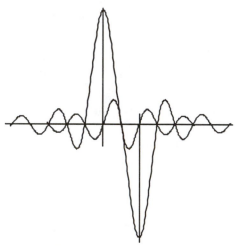

Figure 2.3: Overlapping of Sequential Symbols

From the mathematical form of the symbol waveform given above it is readily shown that the correct interval T between pulses is given by the expression

$$T = 1/(2F)$$

and the rate at which we can transmit symbols over this channel without ambiguity is equal to 2F per second. This is one of the classic results in digital transmission theory and very easily derived, as we have seen.

So far this argument has centered around transmitting only binary numbers where each symbol needs to take on only two distinguishing levels at its peak, positive and negative. However, the situation is essentially the same if the transmitted pulses were to take on a number of discrete levels. The requirement is still that the peak of each waveform, in time, falls on the zero crossings of all the others. Thus if each pulse could be one of eight discrete levels, each symbol could convey three bits of data and the data rate would be three times the symbol rate given above. In general, if N levels are allowed, the data rate is $\log_2 N$ times the symbol rate. The task of the receiver is to distinguish which of the N levels was actually sent.

This model, as presented so far, contains a hidden assumption that does not accord with normal practice. This is the condition that the data to be transmitted arrive at the input to the channel in the form of very narrow voltage spikes, or impulses, so that each excites the impulse response of the channel at the output in the receiver. In practice, of course, the data are more likely to be presented as a stepped rectangular waveform with the steps having the full duration T defined above as the time between successive symbols. The major consequence is that, whereas data in the form of impulses would present a flat frequency spectrum at the input to the channel, real data present a falling frequency characteristic in accordance with the well-known $\sin(\omega T)/\omega$ Fourier transform of a rectangular waveform. Practical systems therefore include a preequalizer in front of the input filter of the channel to compensate for this frequency characteristic. The situation is illustrated in Figure 2.4. The pre-equalizer ensures that the time response of the overall transmission channel to rectangular input waveforms retains the symbol shape described above; i.e. being equal to the ideal impulse response with zero crossings at the required time intervals. All practical data transmitters will be found to contain preequalizers of this nature.

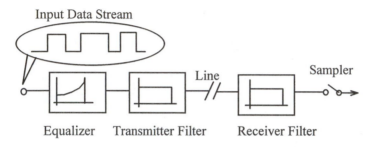

Figure 2.4: Pre-equalization for Rectangular Data Pulses

Spectrum Shaping: The Nyquist Criterion

While the pulse shape in the elementary system described so far is satisfactory to avoid intersymbol interference, the problem remains that the required filter characteristics are idealized "box car" or "brick wall" functions and not realizable in practice. A method is required that will permit the rolling off of the bandpass shape at its edge, and at the same time preserve the zero crossings of the symbol waveforms at the receiver sampling intervals at multiples of time T as described.

It can be shown that the desired symbol zero crossings are preserved if the frequency characteristic is rolled off with some function that is antisymmetric around the original cut-off frequency F, as illustrated in Figure 2.5. This is known as the Nyquist criterion. Apart from being symmetric about F the shape of the rolloff is fairly arbitrary, except that it should be readily approximated by realizable filters and that it should allow the tails of the symbol waveform to decay fairly quickly to reduce the number of symbols contributing to sampling errors when there is a sample timing error in the receiver. In practice, a rolloff approximating a sinusoidal shape is often adopted and referred to by the term "raised cosine" as the resulting form of the frequency response, say H(f), can be written:

$H(f) = 1$ for $0 < f < (1 - a).F$
$H(f) = \frac{1}{2} + \frac{1}{2}\cos[\pi(F - f)/2]$ for $(1 - a).F < f < (1 + a).F$
$H(f) = 0$ for $f > (1 + a).F$

where a is some fraction of the ideal bandwidth F. Large values of a tend to give sharper symbol shapes, easing the matter of synchronizing the sample timing at the receiver, but of course demanding more bandwidth. In broadcasting systems the latter tends to be the more critical and values of a few percent are found.

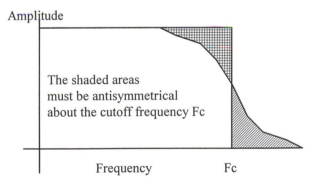

Amplitude

The shaded areas
must be antisymmetrical
about the cutoff frequency Fc

Frequency   Fc

Figure 2.5: Nyquist Shaping of the Transmission Channel

15

Matched Filtering

The necessary frequency characteristic of a transmission channel carrying baseband digital signals is now evident. In practice, this characteristic will be realized by sharp filters at the transmitter and receiver. The filter at the transmitter is necessary to limit the bandwidth of the transmissions, where excessive bandwidth would constitute spurious radiation in radio and broadcasting applications. That at the receiver is required to limit the bandwidth of noise entering the receiver to that strictly necessary, and hence to maximize the signal-to-noise ratio. Both together must constitute the overall desired channel filter response. The question arises, therefore, as to how this filter response should be apportioned between the transmitter and receiver filters.

The result is simply stated but difficult to prove. The filters should be identical, each contributing, in a sense, the square root of the desired overall channel characteristic. This case maximizes the ratio of the midpoint amplitude of the symbol waveform, at the receiver sampling instance, to the level of noise admitted to the receiver. Such filters are said to be matched. It follows from this also that the term "root raised cosine" often appears in the description of transmitter and receiver filters, since the desired channel characteristic is frequently of the raised cosine type described above.

Noise and Error Probability

Consider a system intended to transmit symbols with some number N of defined amplitudes and therefore conveying $\log_2 N$ bits per symbol. With the channel characteristics designed as above, and if the channel were noiseless, the midpoint amplitudes of the symbol waveforms must lie at N clearly defined voltage levels at the receiver sampling instances. Noise adds to these levels in the form of a spurious voltage. Accordingly, the sample values in the receiver exhibit a scatter about their true levels. The task of the receiver is to decide, given a sample value, which symbol amplitude was the one most probably sent. The situation is shown diagrammatically in Figure 2.6. Clearly, the most probable value is that which lies closest to the sampled voltage. Thus, the receiver has decision thresholds at well-defined intervals as the diagram shows.

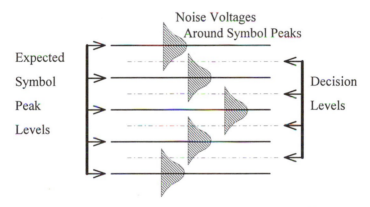

Figure 2.6: Added Noise and Decision Levels

In Figure 2.6, the decision thresholds are shown spaced by a voltage V and the noise voltage is shown as having a statistical distribution with a small standard deviation. The probability that the noise voltage will exceed the value V/2, and hence cause an error, is readily calculated by normal statistical techniques. As V is closely related to the received signal level, and the noise voltage distribution to the noise power admitted by the receiver, the probability of error can be expressed directly as a function of the signal-to-noise ratio. Figure 2.7 shows the typical form of this relationship, in this case for a binary baseband system where the received symbols would have only the two values +/-V.

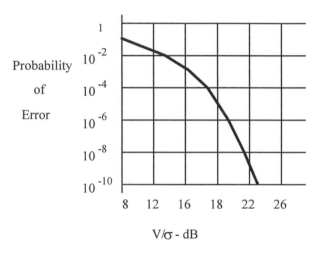

Figure 2.7: Error Probability versus Noise Level

For broadcasting purposes the assumption of steady noise statistics is a little oversimplified, although it may be useful as a comparative measure of the robustness of competing modulation systems. The character of the noise is more often in the form of impulses or spikes from power lines and ignition systems or in sporadic interference from distant transmitters. Hence errors tend to occur in bursts, rather than randomly, and all broadcasting systems include methods of handling such events as we shall discuss under the topics of radio propagation and forward error control.

The Principles of Modulation Methods

Over the years there have been many clever methods for impressing digital information onto an RF carrier, including amplitude phase and frequency modulation. Of these, only amplitude modulation remains important in broadcasting (and in a number of other applications as well). It comes in two guises, linear amplitude modulation, which encompasses vestigial sideband transmission (VSB), and quadrature amplitude modulation, or QAM, which has application in both terrestrial and satellite systems.

In the most general terms, an RF carrier can be considered as comprising a cosine component, written as $\cos(2\pi ft)$, and a sine component, $\sin(2\pi ft)$, often referred to as the in-phase and quadrature component respectively. These can be linearly modulated with separate baseband signals, say $I(t)$ and $Q(t)$, which may or may not be related. Thus, the general modulated signal, say $s(t)$ can be expressed as:

$$s(t) = I(t) \cos(2\pi ft) + Q(t) \sin(2\pi ft)$$

The concept is shown on a voltage phasor diagram in Figure 2.8. Figure 2.8 illustrates the convention by which $I(t)$ and $Q(t)$ are also called the in-phase and quadrature modulation components respectively.

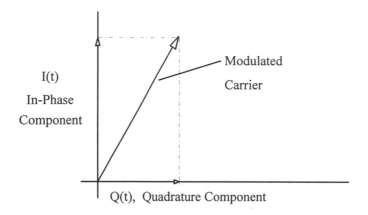

Figure 2.8: In-Phase and Quadrature Modulation

In linear amplitude modulation only the first term is applied, and I(t), for example, would be the stream of symbols from the baseband system described above. The Nyquist spectrum shaping and matched filtering can be applied at baseband or IF. As in analog TV transmission, the occupied RF bandwidth is actually twice what is strictly required, and in the ATSC system using this technique, VSB filtering is applied at the transmitter in a very similar manner to that applying in the analog world.

In QAM systems, I(t) and Q(t) are both streams of baseband symbols like those discussed above, and subject to the same requirements for Nyquist spectrum shaping and matched filtering. However, in this case I(t) and Q(t) are both derived from the one input stream and strictly related so that, at the receiver sampling instances, the modulated carrier can take on only a finite set of values over the in-phase and quadrature domains. These values can be plotted in a diagram where the horizontal axis represents the in-phase component and the vertical axis the quadrature component, both at the receiver sampling instance. These are known as "constellation" diagrams. Figure 2.9.a shows an example where I(t) and Q(t) can each take on four levels, and the resulting carrier may occupy 16 different positions. Not unnaturally, this is called 16 QAM. Figure 2.9.b shows the diagram for 4 QAM. For obvious reasons this is usually, but inaccurately, referred to as QPSK, standing for quadrature phase shift keying. Constellation diagrams can of course, encompass linear amplitude modulation, and that for 8 levels appears in Figure 2.9.c.

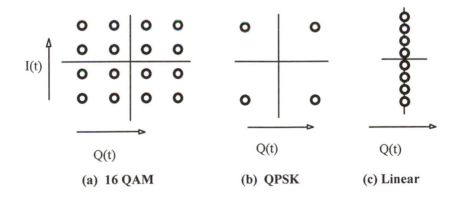

(a) 16 QAM   (b) QPSK   (c) Linear

Figure 2.9: Constellation Diagrams

Constellation diagrams are useful for describing system performance and testing. Thus if noise and distortion are imposed onto the received signal, the constellation positions that the carrier can occupy become spread out and blurred, as shown for 16 QAM in Figure 2.10. The receiver's task at any sampling instance is to determine which of the constellation positions was the one actually transmitted. It usually does this by choosing the one that would lie closest to the position of that being received. Obviously, if the spreading of the received signals around the allowed positions is such that overlap occurs, there is a significant probability that an error will occur in this decision.

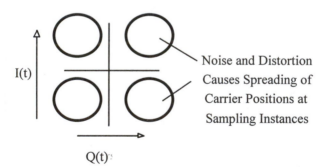

Figure 2.10: Noise and Distortion on Constellation Diagram

On the constellation diagram, one could draw a small line from a correct clean constellation position to the position of a symbol actually received. This would constitute a small vector on the diagram, and a statistical summation of these vectors over a large run of symbols gives a measure of how well or badly the system is performing. This is the basis of the "error vector method" of testing, to be described in more detail in Chapter 11.

Chapter 3

The Effects of Radio Propagation

As a transmission medium the atmosphere is far from perfect. Three effects are particularly unfriendly to digital transmission. These are signal fading, multipath propagation, and impulse noise. In this chapter we look at each in turn and show how they have influenced the design of digital broadcasting systems.

Signal Fading

Those who live on the outer fringes of a broadcasting service area know that the reception quality can be quite variable. It can vary from location to location, in that where one household might enjoy quite good reception neighbors a few streets away might have to use elaborate antenna systems to obtain even a just reasonable service. The quality can also vary with time and is particularly influenced by weather conditions.

The location variability is generally caused by what is known as *clutter loss*. The name reflects the cause, which is quite literally the blocking of the transmission path by the clutter of buildings, trees, and local hills to be found in the vicinity. This loss tends to be fairly stable but is influenced particularly by the rain wetting trees and the surfaces of buildings.

The principal cause of the variability in time is the effect of atmospheric refraction over the transmission path. Consider a typical situation shown in Figure 3.1.a. This shows a ray line from a transmitter to a domestic receiving antenna. The ray line is drawn as a straight line to show the situation when the atmosphere is uniform. Notice that it clears the obstruction drawn at about midpath.

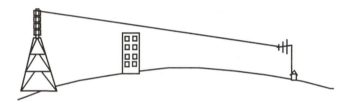

Figure 3.1.a: Clear Ray Line, Normal Refraction

Now let us suppose the refractive index of the atmosphere changes with height above the ground. This happens whenever there is a temperature gradient, as the refractive index of air is temperature dependent. Recall also that the refractive index governs the speed of propagation of electromagnetic waves, light or radio signals, through a medium. Hence, if the variation in refractive index causes, for example, the lower edge of the wave path to have a have a greater velocity than the upper edge, there will be a tendency for the wave to curve upward. The effect is that a ray line between the transmitting and receiving antennas must take a concave upward path as shown in Figure 3.1.b. Notice now that the ray path could now be partially blocked by a midpath obstruction as shown in the figure.

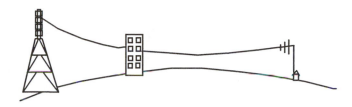

Figure 3.1.b: Ray Path Curved by Refraction

To take this example one step further, Figure 3.1.c shows Figure 3.1.b redrawn and distorted in such a way that the ray path appears straight. The distortion has the effect of increasing the apparent curvature of the earth, or equivalently, making the radius of the earth seem smaller than it actually is. Hence, the literature on propagation often refers to a "k factor," where k is the number by which the radius of the earth would have to be multiplied in diagrams like Figure 3.1.c to make the ray path appear straight.

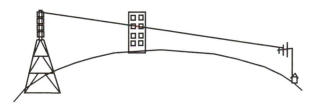

Figure 3.1.c: Earth Curvature Shown Increased

Normally k has a value of around 1.4 and varies over a range of approximately 0.8 to 2 in most climates. We see that as k changes typical ray paths will become more or less blocked by obstructions or even by the earth's bulge, and this largely governs the extent to which the signal strength varies over any propagation path.

This explanation will be seen as a little oversimplified, as it must be obvious that the energy propagating from the transmitting to the receiving antennas cannot be confined to the single line we have called the ray line. In fact, the energy occupies a volume of space contained within what are known as the Fresnel zones, named after a famous researcher into electromagnetic wave propagation. Anything that penetrates into the Fresnel zones causes some loss of received signal level. In broadcasting, the earth's bulge and ground obstructions always penetrate into the Fresnel zones, and the propagation loss is greater than it would be in free space. Hence we can say that almost any variation in atmospheric conditions, by causing a change in k, will cause the zones to be more or less obstructed and cause a variation in received signal level.

System designers must take into account the variability of signal level both in location and in time. To assist them both the Federal Communications Commission (FCC) and the International Telecommunications Union (ITU) have published graphs that show the signal levels to be expected in typical situations. The curves take the form shown in Figure 3.2, which is a very simplified version.

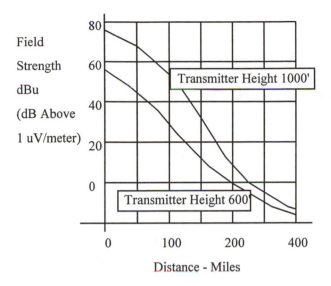

Figure 3.2: Typical F(L,T) Curves

These curves plot the expected level of field strength against distance for one kilowatt radiated from various transmitting antenna heights. They are universally known as the F(L,T) curves where F denotes field strength, L is the percentage of locations, and T is the percentage of time for which the field strength will be greater than the value plotted. Thus the F(50,90) curves show the field strengths that will be exceeded for 50% of locations and 90% of the time at various distances in typical terrain. Correction factors are available to adjust the values for differing terrain characteristics, for example for propagation over water or hilly districts. The curves have been derived from theoretical analyses adjusted to take into account many years of measurements in the field.

In analog system design the F(50,50) curves are the ones normally used to determine the necessary transmitted power level. However, in setting the parameters for the United States digital TV plan, the FCC used the F(50,90) curves, and many people argue that digital system design should be based on the much more severe requirement of F(95,95). To see why, we must look at the different failure mechanisms of analog and digital systems under the conditions of signal fading.

When signal fading occurs in an analog system the net effect is to increase the level of video or audio noise. As fades deepen, the picture or audio quality gradually deteriorates, and even in relatively deep fades the picture remains watchable and the audio understandable. Analog systems are therefore said to fail "gracefully." Most people find this tolerable as long as the quality is good most of the time.

On the other hand digital systems fail catastrophically. Recall that each pixel value or sampled audio level is converted to a binary number and transmitted as a sequence of bytes comprising a binary word. Thus if the error rate is small, most numbers will be recovered correctly at the receiver, with the occasional one being interpreted incorrectly. In TV, the picture will be essentially perfect over the screen with only the odd dot showing from a pixel with incorrect brightness or color. In audio, only an occasional click will be perceived. However all systems contain forward error correction, and at moderate error rates even these artifacts will not be present and the perceived service quality will remain perfect.

We shall see in Chapter 5 that forward error correction can correct only up to a certain number of errors in each binary word.

For example, say that number is 10. Then, if 11 errors occur the system is unable to make the necessary correction, and if the error is in the most significant bits, the receiver is likely to interpret the number as totally different from that transmitted. Hence, the decoded output level will also be totally different from that transmitted. Note that this is much worse than what happens with added noise in analog systems. That at least leaves the output level usually somewhere close to the value transmitted. At the point where the forward error correction system is about to fail, the system will be operating at a significant underlying error rate. From the graph of error rate versus signal level in Figure 2.7, we see that if the error rate is significant any small reduction in signal level will cause a great increase in that error rate. Hence, as the fading RF input signal falls below the level at which the forward error correction loses control, the picture and audio quality collapses dramatically. Figure 3.3 contrasts the manner in which analog and digital systems fail as the input signal level declines.

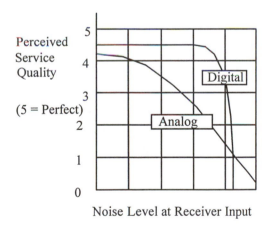

Figure 3.3: Contrasting Failure Modes

This effect explains why the signal level for digital systems must stay above some critical level for a much greater proportion of locations and times than is necessary for analog. Some viewers may put up with a noisy picture for 50% of the time, but they certainly would not accept having an unusable service to anything like that extent.

Multipath Propagation

Multipath describes the situation whereby a signal reaches a receiver by more than one propagation path from the transmitter. Normally, the main signal reaches the receiver by the shortest direct path, and a second signal arrives later, having taken an indirect path by virtue of being reflected off some large object situated away from the main path. Some peculiar atmospheric conditions can also cause two curved direct paths to come into existence. The delayed signals are often described as echoes. They cause ghosting in analog TV and can give rise to severe distortion in FM services, as anyone can attest who has driven through the centers of large cities where many reflections from buildings are often present.

Digital systems are very sensitive to multipath. In Chapter 2 we showed the form that the transmitted symbol waveform must take to ensure reliable detection at the receiver. In particular, the waveform must have a large central peak surrounded by zero crossings at specified instances in time before and after that peak. The signaling rate is arranged so that the zero crossings of leading and trailing symbols fall at the midpoint in time of the central peak of any given symbol. Thus, leading and trailing symbols do not interfere with this central peak, and its level can be accurately determined at the receiver.

Now consider what happens when an echo is present. The peak of a symbol in the echo may fall effectively anywhere, including close to the central peak of some symbol sent earlier. It may thus cause the detected level of the wanted symbol to change considerably, possibly leading to an error and certainly leading to an error caused by a lower level of noise than would otherwise be the case.

To be specific, let us consider a system in which the symbol amplitudes may take eight discrete values as in the ATSC system used in the United States. The normalized symbol amplitudes will be +/-7, +/-5, +/-3, and +/-1 V and the receiver decision levels will be spaced ½ V from each of these. Now suppose we transmit a +7 V symbol followed by a +1 V symbol, but because of multipath, a delayed and attenuated echo of the 7 V symbol comes to overlay the 1 V symbol. This will definitely cause an error if the echo magnitude exceeds ½ V. Hence the echo attenuation must be better than 14 times or 23 dB. More particularly if the echo is ¼ V it will reduce the signal-to-noise ratio by 6 dB. This is an echo attenuation of 30 dB, comparable to the level of ghosting tolerable in analog TV. Given the failure characteristics of digital systems described, it represents an intolerable loss in service area or reliability.

Note that the time between symbols in the ATSC system is 93 nanoseconds, and typical echo delays in urban environments are some tens of microseconds. Hence, the situation described above can readily arise, and all digital terrestrial broadcasting systems require some method of combating multipath. This is not so true for satellite systems as the receiving antenna beamwidths are usually sharp enough to reject reflected signals.

The two major modulation standards emerging for terrestrial transmission are the ATSC eight level vestigial sideband (8VSB) system and coded orthogonal frequency division multiplexing (COFDM). The methods of combating multipath in these will be dealt with in Chapters 8 and 9. Briefly, however, the ATSC system uses an active equalizer which inserts echoes at the receiver in the inverse sense to those coming in on the transmission path and thus brings about their cancellation. The transmitted signal contains a defined "training signal" for the equalizer. By subtracting a local version of this signal from that received, the equalizer can determine the character of the echo signals present. COFDM uses very long symbols and adopts a "guard band" in time, during which it simply ignores the incoming signal. The guard band is long enough for the echoes of leading and trailing symbols to have died out for typical urban multipath delays. Hence we see that combating multipath is a very significant feature of practical systems. One can make the definitive statement that without it no digital terrestrial system would work.

Impulse Noise

Treatments on digital transmission in communications systems usually focus on its immunity to steady electrical noise, called thermal noise after the mechanism causing it, or Gaussian after the famous mathematician who gave his name to the statistical distribution characterizing it. This noise has the advantage of being well understood and fairly easily analyzed mathematically. In broadcasting, however, we have to be concerned with the more insidious impulse noise that pervades most modern towns and cities.

Impulse noise has many causes, including motor vehicle ignition, electric motor starters, and many industrial processes. A particularly important source is surface leakage across power line insulators, where the noise comes in bursts at the power line frequency. It is often seen on analog TV sets as a band of noise drifting slowly up or down the screen. Its level is largely unpredictable, as is the probability that it will occur in any particular location. Impulse noise is an annoying nuisance in analog transmission, but rarely causes reception to be unusable. It can, however, cause intolerable interruptions to digital transmissions.

Chapter 4 on transport streams, shows that a significant portion of the digital transmission stream is given over to synchronizing signals of some form. All digital receivers have to establish synchronization at a number of levels. All this takes time and, for example, when a digital receiver is first switched on, it may take several hundred milliseconds to establish lock and produce the desired program output. Relocking is also generally required when changing channels. Making this sufficiently fast to avoid annoying viewers has been one of the major challenges in designing domestic receivers.

The transmission of the synchronizing signals is usually made more robust than that of the service content, to ensure that if a fade does occur the synchronization will be preserved even if the picture and audio fail, so that relocking is required as infrequently as possible. However, given the unpredictability of impulse noise there is no guarantee that it will not disrupt the synchronizing signals from time to time.

Impulse noise characteristically comes in short bursts, and when it occurs, it is likely to take out a lengthy string of received symbols. One way of countering its effects then, is to use forward error correction that is effective over a run of several errors. Most forward error correction, however, is effective only for single errors or very short runs of errors. Hence, the second technique is to shuffle the order in which the payload bits are transmitted and to reverse this shuffling at the receiver. The shuffling is made as random as possible. Hence if impulse noise strikes, the errors will be contiguous in the received stream, but after the bit stream is reverse-shuffled, the errors become randomly scattered through the data stream, and their spacing depends on the number of bits involved in each shuffled group. The randomly distributed errors are then readily corrected by the forward error correction circuits in the receiver.

This technique greatly improves the robustness of the transmissions to impulse noise. It is defined for all current terrestrial system standards.

Chapter 4

Transport Streams

This chapter shows that even the most elementary digital transmission systems require the transmitted information to be organized and identified in some way. When the system is carrying more complex systems, such as video, several audio channels, and perhaps data channels as well, it becomes imperative that there is some way of associating the bits with the correct service. This is the process of multiplexing and demultiplexing, by means of which bit streams for a number of services are combined for transmission over a single link and then separated for delivery to the appropriate output device.

In program distribution, before they reach the transmitter, the broadcasting services may pass through a variety of transmission networks, including optical fiber, microwave systems, and satellites in which they may be multiplexed with and demultiplexed from other digital streams any number of times. Hence, the chosen method of identifying program segments in the broadcasting stream must be compatible with national and international standards for multiplexing broadband services. The two concepts we will introduce here are those of packetized elementary streams and of transport streams.

The Need for Packets

The most elementary digital transmission system that one might conceive of is that for transmitting sampled values of a mono audio service over a dedicated channel. For simplicity, let us suppose that each sample must be transmitted as four binary bits and hence is approximated by one of the 15 values +/-7, +/-6, . . , 0. If some sequence of values were 6, 3, 2, 1, 5, 7, 2, 5, 3, say, the transmitted bit stream would be 0110, 0011, 0010, 0001, 0101, 0111, 0010, 00101, 0011. Now, let us take out those commas, and the bit stream becomes 011000110010000101010-1110010001010011. What is the receiver to make of that? How can it tell where one number ends and the next one starts? Obviously, we must put back those commas. However we don't need to put back all of them. We could, for example, send the numbers of groups of three in which case the bit stream would become 011000110010, 000101010111, 0010001010011. The comma still turns up regularly and by finding it in the bit stream the receiver can remain synchronized to the transmitted bit stream and determine the starting position of each transmitted number.

At this stage we need some way of transmitting the commas, which obviously must be an easily detected string of bits. In this particular example the sequence 1111 may suffice, as being at the extreme range of our audio sample values, it turns up rarely in the natural range of numbers to be transmitted. However, if it is inserted after every third number like the comma above, the receiver can be designed to seek it out at regular times.

In the terminology normally used for digital transmission, each group of four digits would be referred to as a byte, and the sequence 1111 as the synch byte, where synch is short for "synchronizing." To be strictly accurate, the term *byte* is normally applied to strings of eight bits, but we use four in this example to keep the discussion easier.

The normal method of synchronizing the receiver timing to the synch bytes is by means of circuitry known as *phase locked loops*. The principle is shown in Figure 4.1.

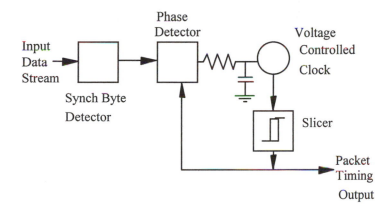

Figure 4.1: Synch Byte Timing Recovery

Broadly, a synch byte detector produces a pulse whenever the synch byte is detected. Its output signal consists of a string of pulses spaced at the regular periods at which the synch pulses are inserted at the transmitter, but with the odd random pulse occurring when a data byte happens to match the synch byte. At the same time a voltage-controlled clock produces a regular train of pulses whose period closely approximates the expected period of the synch bytes. As the name implies the voltage-controlled clock is an oscillator whose frequency can be changed by applying a control voltage. The two trains of pulses enter a phase comparator for which the output voltage is proportional to the phase, or timing difference, between them. It is easy to design a detector that is sensitive to the regular synch byte pulses and ignores the irregular false pulses due to data bytes. The voltage out of the phase detector is fed back to control the frequency of the voltage-controlled clock. This serves to pull the frequency so as to correct the detected phase error between the clock pulses and the synch byte pulses, so that the clock is said to be locked to the synch bytes. This serves to define the master clock timing in the receiver. The timing for all byte detection and interpretation is taken from this clock.

Packetized Elementary Streams

We have just seen that even the most elementary data stream needs to be organized into packets separated by easily recognized synch bytes. However, once the packets are recognized, there is no need for all the bytes in them to actually be numbers representing sample values (audio samples in this example). In practice the number of bytes in a package would be vastly greater than the three used in the example and a few of them could be dedicated to conveying information to the receiver.

To persist with this elementary example, suppose the basic audio transmission system is to be extended to a stereo mode. One possibility is to make the packages carry alternately samples from the left and right channels. We need to tell the receiver which channel each package belongs to, and this might be done by making the first of the three bytes in each package either of two special bytes representing left or right. In the receiver, simple circuitry looks at the first byte and switches the output to the left or right channel as appropriate. Importantly, the receiver is designed to use this control byte for some specific purpose and not as a sample value. In reality, there may be a number of these control bytes in each package. Some current practices are described in the chapters on video and audio compression.

Packaged streams like this are found typically in the video and audio streams in a broadcasting multiplex. Some bytes in each packet may be given over to carrying program associated data, often abbreviated to PAD, which conveys data like song titles or the names of artists. An important component in each package is commonly a time stamp, which shows where that package is related in time to all the others in the multiplex. The multiplex may also support a data stream that may be used to carry almost any information. Each of the streams which thus make up the multiplex is known as a packetized elementary stream or PES for short. Each PES typically carries a single compressed video channel, a multichannel audio service, or data in some format.

Transport Streams

In general, each of the PESs that come together to form a multiplex may contain packets of different lengths, as befits the applications concerned (e.g., video, audio, and data in broadcasting). It is desirable, of course, that the method of multiplexing all these services together results in a bit stream that is itself packetized into fixed-length packages with its own unique synch byte like that described in our elementary example.

41

The way this multiplexing is carried out is illustrated in Figure 4.2. The elementary streams enter the multiplexer in parallel. The stream leaving the multiplexer consists of a sequence of packets of some fixed number of bytes.

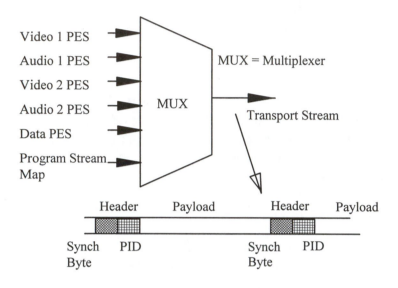

Figure 4.2: Multiplexing of Elementary Streams into the Transport Stream

Each packet has a synch byte and a short header containing information for the receiver. The header is followed by the "payload bytes," a sequence of bytes taken from one of the elementary streams. Part of the header is a program identifier, or PID, which identifies the PES from which the payload bytes have come. One of the PESs, called an elementary stream map, carries data indicating the relationships between the other elementary streams and how they may go together to make up programs. For the situation in Figure 4.2, for example, the elementary stream map would indicate that video 1 and audio 1 make up one TV program.

The lengths of the packets in the individual PESs need not be directly related to the length of the packets in the transport stream. If, say, the PES packet contains enough bytes to fill a whole number plus some fraction of transport stream packets, that part of the final transport stream package left vacant is filled with dummy data. This process is often known as bit stuffing. It introduces some inefficiency in that more than the strictly necessary number of bytes has to be transmitted. However, this is nearly always outweighed by the convenience of having fixed-length transport stream packets and of being able to deal with PES packets of any length.

In the ATSC and MPEG realizations of transport streams each packet contains 188 eight-bit bytes. The header includes a 13 bit PID that identifies the PES carried, and also contains a program clock reference. This latter is used to compensate for time variations that may occur in transmission, as telecommunications carriers change transmission routes from time to time to optimize their traffic loading.

Apart from simplifying synchronization at the receiver as we have seen above, the use of fixed-length packages in the transport stream brings a number of advantages. Receiver designs can be standardized over terrestrial, cable and satellite systems. Error correction is more easily applied to fixed-length packages. They are easily inserted into higher level multiplexes used by national and international communications carriers. The number of PESs that can be inserted into a transport stream is variable and can be adjusted to suit the bandwidth available. The content of each PES is immaterial, and there is much commonality between TV, audio, data, and even communications applications. In broadcasting, each transport stream could carry a variable number of programs varying over high definition TV to VCR quality depending on the bit rate assigned to each.

The level of packaging just described is appropriate to transmitting broadcast services to the general public. The generation of higher-level multiplexes is, however, very much an extension of the process used to form the initial transport stream. Naturally the identification of the contents of the higher order packages becomes more and more complex, and may extend to carrying addresses as each package may have a different destination. In the end, however, it all comes back to putting the commas into a data stream to identify which byte is which.

Chapter 5

Forward Error Correction

Most digital transmission systems incorporate some form of error detection and correction. It is often a vital component of the system. For example, imagine how unacceptable it would be if downloaded computer programs always contained some undetected errors. In broadcasting, it is not quite that critical, but even the occasional error could have bigger than expected consequences. In Chapter 7 on Moving Picture Expert Group (MPEG) picture coding it can be seen that one error might propagate over several picture frames, for example. Error control also allows the system to operate at lower signal strengths than might otherwise be the case, and hence contributes to minimizing the transmitter power required for a given quality of service. It also makes a less obvious contribution to the realizability of broadcasting services. For example, it permits the clipping of rare but very high peak voltages in transmitter power amplifiers by correcting the odd error introduced by this event.

This chapter explains the principles behind the most common types of error control used in current broadcasting transmission systems. It introduces block codes and convolutional or trellis codes. Broadcasting standards often use the terms *inner codes* and *outer codes*. The words really refer to where they are found in the transmission system. Thus, the inner codes operate just before the signal leaves the transmitter and just after the receiver demodulator. Outer codes tend to be applied close to the extreme input and output ends of the transmission chain. Inner codes are usually convolutional in nature, with an optimum performance under conditions of steady noise interference. Block codes are usually more effective for bursts of errors and are used for outer codes.

It will be useful in this chapter to use the concept of a binary word. A binary word is merely a string of binary symbols, or bits, that can be any defined length. In practical error control systems the binary words may be hundreds of bits long, but they will be much shorter for the examples that appear in this chapter.

Parity Testing

The most common form of error detection is the use of a parity bit. *Parity* means adding an extra one or zero to the end of a binary word so that the total number of ones contained in the new word is either *odd* or *even*. If odd, we are said to have elected to use odd parity. It is even parity if we elect to make the total number of ones even. To take an example that used to be common in computers, suppose we adopt odd parity by adding a ninth bit to each eight-bit byte so that the total number of ones is odd. Thus we would add a one to the byte 01010000 to make the nine-bit word 010100001. Now if one error were to occur in that word a test on the parity would fail, inasmuch as the total number of ones would become even.

Notice that the parity test cannot identify the bit position at which the error occurred. Hence, in can detect the error, but cannot correct it. Systems using simple parity tests usually employ a strategy called ARQ, or automatic repeat request, whereby once errors are detected the receiver sends a request to the transmitter to repeat the data concerned. This is not possible in broadcasting, of course, and error control must take the form of FEC, or forward error correction, using codes capable of both detecting and correcting errors.

Block Codes

The simplest type of error correction employs an extension of the parity concept. For simplicity, we will look at how this would be applied to very short binary words, although in practice, of course, the words would be much longer.

Suppose we were to break up the data to be transmitted into four-bit words and arrange them in a four-by-four matrix (or block) as shown in Figure 5.1. Call these the message bits, designated as "m" in that figure. Now at the end of each row and at the bottom of each column, add a parity bit so that the word in each row and column has even (or odd) parity. These are designated "c" in the figure, as they are commonly referred to as "check bits" in the literature. We now have a 25-bit block containing both message and check bits. This can be transmitted by some method as a 25-bit word.

```
m m m m c
m m m m c
m m m m c
m m m m c
c c c c c
```

Figure 5.1: Block of Message and Check Bits

Imagine that at the receiver this block is reassembled into the same form it had in Figure 5.1. If an error has occurred the parity check will fail on both the row and the column that contain the incorrect bit. Thus, this scheme not only detects an error, but also indicates where it lies. It is simply then necessary to invert the bit in that position to correct the error.

To indicate where the error lies, the decoder has to generate some output word. In its simplest form it might well be a 16-bit word with zeros in all positions except that in which the error occurs, and with that position containing a one. Output words of this nature are known as *syndromes*, derived from the medical usage of the word where the syndrome is the collection of symptoms for a disease.

Notice that if two errors occurred, two rows and two columns would fail the parity check, giving four possible locations for the two errors. Hence, this particular method of coding can correct only one error in each block. In fact, almost the only virtue of this method is its simplicity. There are, as we shall see, much more complex methods of generating the check bits to obtain a better result.

The transmission of the 25-bit word can take almost any form desired. A systematic method would be to transmit the 16 message bits followed by the nine check bits. This overall technique would then, not surprisingly, qualify for the name of a block systematic code.

Block systematic codes usually go under the designation Name(N,M), where N is the total number of bits in a word and M the number of message bits. Thus this example would be designated Name(25,16). The name is often derived from the initials of the inventor of the particular code.

Haming Distance

We have seen that the simple coding scheme above can correct only one error in each word. Practical codes, of course, can correct many errors. One method of appreciating how the more complex are structured is through the concept of the Haming distance, named after a very famous pioneer of coding theory.

The Haming distance between two words is the number of positions in which the two contain different digits. For example, the Haming distance between the two five-bit words 01101 and 11001 is two, since they differ in the first and third digits.

Notice in the simple code above that with 25 bits it is possible to form a total of 2^{25} or nearly 33.6 million different words. Of these, only 2^{16} or about 65000 will ever be transmitted, since only this many different message sequences are possible with

16 message bits. This means that 99.8% of the possible 25-bit words are never transmitted, and if one should be detected out of the receiver it will contain one or more errors. However, if its Haming distance from one of the permitted words is smaller than that from some other permitted word, then more probably the former was actually transmitted.

This last statement relies on the fact that one error in a word is much more common than two, two errors much more common than three, and so on. This method of choosing the word most likely to have been sent is an example of maximum likelihood decoding, of which we will meet another example later. It is a very common strategy in coding schemes.

Suppose now that the minimum Haming distance between any of the permitted words is three. If a word is transmitted and a single-bit error occurs, the erroneous received word would lie at a Haming distance of one from that transmitted. However, it must then lie at a distance of two from some other permitted word. In this case the receiver makes the correct decision by choosing the permitted word with the smallest Haming distance from that received. However, if two errors had occurred, it would make the wrong decision, because the erroneous word now lies closer in distance to another permitted word. Thus we see that with code words spaced at Haming distances of three it is possible to correct a one-bit error in any word.

It should now be apparent that to correct two errors per word, the permitted code words should be spaced at Haming distances of five, or for three errors, at distances of seven and so on. The clever part of designing a code is to pick out that set of words which all lie at the appropriate Haming distances and employ only those for the permitted code words.

At face value this would seem to be simple, but in practice it turns out to be a far from trivial problem. To illustrate, it can be shown that in our population of 33.5 million 25-bit code words, the selection of those with Haming distances of five comprises only 409 members. Hence, this is the maximum number of different messages we could transmit with 25-bit words and still expect to correct two errors per word. That is not very efficient. In this argument lies the reason why practical codes comprise very long code words. Only when there is a large number of bits in each word does a significant proportion of the population lie at Haming distances of the order of five to ten.

Cyclic Block Codes

Practical decoding schemes at the receiver are, in principle, not very different from that implied by the matrixing process in Figure 5.1. In effect the receiver has to form a matrix from the received word and calculate various relationships over the message and check bits. Alternatively, it might contain a look-up table of permitted code words to determine that with the closest Haming distance to the one received. We have seen, however, that for the correction of multiple errors, the code words must be very long. Hence, either the computational load or the memory requirements in the receiver can become enormous.

A way around this difficulty has been found by using a group of codes known as cyclic block codes. In effect the only permitted words are those which can be generated by shift registers with feedback from output to input. These registers produce very long trains of bits that repeat after some interval (hence the term *cyclic*). It turns out that with this constraint all the necessary operations like calculating the check bits and forming the syndrome in the receiver can be performed by feedback shift registers which are both fast and simple to implement. Almost all modern block codes are of this type.

The most common of the cyclic block codes are the BCH codes where BCH stands for the names of their inventors, Bose, Chaudhuri, and Hocquenghem. The most famous are the subsets developed by Reed and Solomon. Thus in the ATSC digital TV standard we find the R-S(207,187) code. Other variations of RS codes turn up in European Standards for TV and radio broadcasting.

Convolutional Codes

As seen, block codes take a group of bits, operate on them to form the code word, dispatch that down the transmission channel, and then take the next group of bits for the same treatment. This might be described in industrial parlance as a batch process. In contrast, convolutional coding is more akin to a continual flow process. Convolutional coders also operate on a group of bits, but the group flows continually through the coder and changes only one bit at a time as the data bits arrive. Thus the effect of any one bit ripples through the output stream for as long as that bit remains in the coder. This ripple-through effect is similar to a mathematical process known as *convolution*, literally a rolling together of the effects of many causes. As we shall see, diagrams that illustrate the process look similar in form to garden trellises, and consequently this coding technique is very often referred to as trellis coding.

Convolutional coders do not output words, as do block coders, but rather a continual stream of bits. However, the bit stream cannot take on all possible sequences of bits. Only certain bit sequences are allowed and those transmitted are determined by the sequence of bits in the incoming data stream. It follows that, since the incoming data stream can encompass all possible sequences, but the outer stream only a subset of that, then the output bit rate must be greater than the input rate. The relationship between these bit streams defines the "rate" of the coding process. Thus a rate 2/3 implies that the message bit rate is two thirds of the coded bit rate and so on.

For the purposes of illustration we will adopt the simple rate one-half coder shown in Figure 5.2. Note that each of the outputs, labeled X and Y, depends on both the incoming message bit and also on the contents of the shift register. The contents of the shift register is conventionally referred to as the "state" of the coder, and one sometimes sees references to these devices as "state machines." In this particular case the coder can be in any one of the four states 00, 01, 10, or 11. For each input bit, the output switch arranges for first X and then Y to be output. Thus, two bits are output for each input bit, giving the coder its rate one-half property.

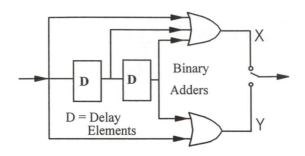

Figure 5.2: Rate One-Half Coder

Figure 5.3 is the trellis diagram for the coder. On the left-hand side are the four possible states of the coder. It will be in any one of these when an input bit arrives. The state of the coder will then change to a new state, shown on the right-hand side of the diagram, with that state depending on whether the input bit is a one or a zero.

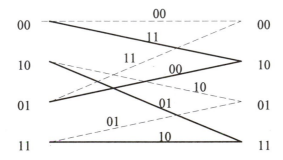

Figure 5.3: Trellis Diagram for Rate One-Half Coder

The possible paths from the left hand to the right hand states are shown by the lines drawn between them, full lines representing a one input and dotted lines a zero input. As stated, the output depends on both the starting state and the input. It is shown by the two-digit byte marked above each of the lines in the figure.

Let us now trace through the sequence of events that occur in the transmitter and the receiver during the transmission of a short string of data bits. To make the explanation simple and the diagrams legible we will confine this to a string of only a few bits and choose the sequence 1010. The situation at the transmitter is shown in Figure 5.4. This indicates the path through the state trellis for the transmitter starting in state 10 and shows that the output sequence would be 01 01 00 10. In this diagram the states are labeled a, b, c, and d for clarity in the following discussion.

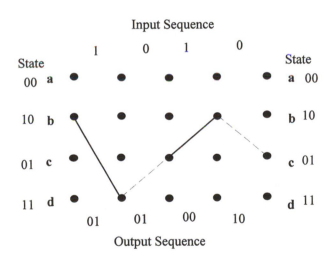

Figure 5.4: Trellis Path in Transmitter

Now let us turn our attention to the receiver. If the receiver is in state 10, the paths through the trellis diagram for all possible sequence are shown in Figure 5.5.a. This diagram is confined to only three shifts as otherwise it becomes unreadable.

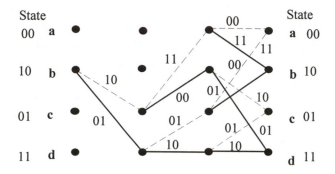

Figure 5.5.a: Possible Paths From State 10

The starting state in the receiver may not be the same as in the transmitter, particularly if there have been errors in the bit stream previously received. Hence, if the receiver were to be in state 00, the possible paths become those shown in Figure 5.5.b. We could draw similar diagrams for the other starting states.

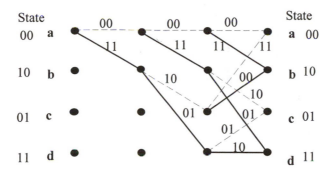

Figure 5.5.b: Possible Paths From State 00

The action at the receiver is to take all possible paths through the trellis diagram and calculate the Haming distance of each of them from that actually received. For example, suppose some path would generate an output sequence 01 11 00 00. Compared to the output sequence in our example, namely 01 01 00 10, this gives a Haming distance of two. Of course, if there were no errors, some path must give a Haming distance of zero. The receiver selects the path with the lowest Haming distance as being the one most likely to have occurred at the transmitter. Here again is an example of maximum likelihood decoding.

A moment's reflection will show that this is no trivial task for the receiver. In principle, it could operate on sequences of any length, as there are no defined blocks of data in this continuous flow process, and at any time it cannot be sure that it is in the same state as the transmitter. The solution lies in adopting a strategy known as the *Viterbi algorithm*, named after its inventor and one of the most famous principles in coding theory.

The Viterbi Algorithm

The key concept in the Viterbi algorithm is that of "survivors." To explain this let us take the first six bits, namely 01 11 10, generated by our test sequence. Let us compare this with the outputs that would be generated by a few of the paths in Figure 5.5. The Haming distance between the output of a particular path and the received sequence is usually called the metric of the path and we shall adopt that in this discussion. Hence, we can now write down the state sequences for a few paths, their output bit streams, and their metrics. Referring to Figure 5.5 we have:

Path bcab Sequence 10 11 11 Metric 5
Path abdc Sequence 11 01 01 Metric 2
Path bcbc Sequence 10 00 01 Metric 4
Path bddc Sequence 01 10 01 Metric 3

Notice that the last two paths start at the same state and finish at the same state. Of the two, however, the last has the smaller metric and is likely to be the more correct if there have been transmission errors. The Viterbi strategy is to discard the path with the larger metric. The remaining path becomes the survivor path linking two given states. For a real and lengthy sequence, this strategy dramatically reduces the number of possible paths that the receiver has to store and compare with the input sequence. One can see this with a cursory examination of Figure 5.5. In principle there would be only one survivor path from each of the starting states on the left to each of the finishing states on the right, no matter how far the diagrams were to be extended.

There are variations on the Viterbi algorithm. For example, in one strategy the receiver follows a path while its metric remains small. If the metric increases beyond some preset limit the receiver backtracks over a few steps and tries a branch different from that previously traversed. These algorithms are comparatively complex and are usually implemented in software on dedicated LSI chips.

Applications

The coding scheme used in the ATSC standard is illustrated in Figure 5.6. Two input bits are coded into the three bits required to define one of the eight symbol amplitudes. This is done by leaving one bit unencoded and passing the other through a simple rate one-half coder. As mentioned in Chapter 8 on ATSC, there are in fact 12 of these encoders and each operates on data bits spaced 12 apart in the payload stream.

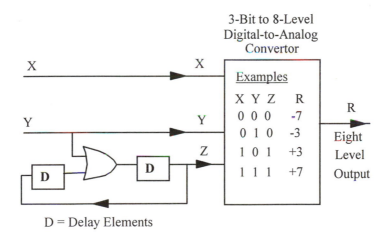

Figure 5.6: Coding Used in ATSC Standard

59

The encoder applied in the European DVB standards is illustrated in Figure 5.7. It is capable of being adapted to various coding rates by controlling the switching between the X and Y output channel. For rate one-half the transmitted sequence is XY for each input bit as in our simple example. For rate three-quarters, the output is the four-bit sequence XYYX for every three bits input.

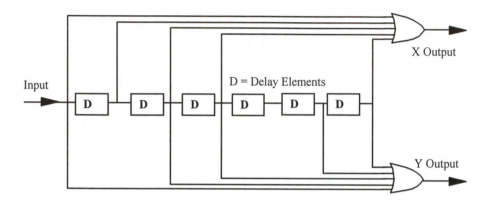

Figure 5.7: Encoder Used in DVB Standards

The ability to change the code rate can be useful. For example, one may choose to encode a TV service as a standard definition picture with, say, rate one-half coding and also as a high definition picture with rate seven-eighths coding in the same service channel. Hence, some viewers will be able to watch the program in its high definition form, but those with poorer reception conditions can fall back on the more robust standard definition service.

Chapter 6

Audio Compression and Coding

The compact audio disk is often cited as the prime example of how adopting digital techniques can bring better and cheaper solutions than analog methods have ever done. Yet despite the fact that the CD almost entirely replaced vinyl records many years ago, digital techniques have been slow to penetrate into transmission systems for broadcast audio, either for radio services themselves or for the audio channels associated with video in TV. One of the reasons for this is that digital systems delivering high audio quality, comparable to the quality of FM radio, have required relatively enormous bit rates prior to the development of the compression techniques described in this chapter.

Essentially, digital audio transmission entails sampling the signal at a certain rate, approximating the sampled values with numbers, and then transmitting that series of numbers. The rate at which the samples must be taken is related to the audio bandwidth by the sampling theorem. Since each sample value must be approximated from a finite range of numbers there is a small error for most samples. The sequence of these small errors in the transmitted data stream gives rise to what is known as quantizing noise in the reproduced signal. This chapter outlines the principles underlying the sampling theorem and quantizing noise, and shows that they both contribute to the high bit rates required for good-quality audio.

In recent years, a series of experiments has shown that the human auditory system does not necessarily use all the detailed structure of the audio signal that these high bit rates are capable of conveying. In other words, much of the information is actually redundant and can be discarded with no effect on the perceived sound quality. The redundant information is removed by using mathematical models of human hearing to control the processes of sampling and quantization. These are known as psychoacoustic models. In broadcasting the most common applications are found in MPEG and Dolby AC-3 compression systems, which are briefly described in this chapter.

Bandwidth and Sampling

In its simplest form the sampling theorem states that for accurate reproduction a signal of bandwidth B hertz must be sampled at a minimum rate of 2B per second. To demonstrate this, let us consider the simple sampled transmission system shown in Figure 6.1.

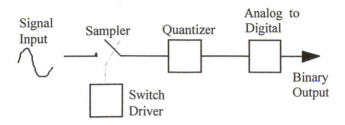

Figure 6.1: Sampled Signal Transmission System

The sampler on the left-hand side takes samples at time intervals T. The samples are quantized and sent to the receiver through some transmission system as a sequence of numbers. The numbers might be transmitted using any of the methods described in Chapter 2. At the receiver the sequence of numbers is converted back into a stream of pulses that have the same quantified amplitudes as they had at the transmitter. These pulses are then passed through a low-pass filter of bandwidth B to approximate the original signal. This is a simple, but accurate, model of sampled digital transmission.

Now instead of thinking of the signal being sampled at the transmitter, consider instead that the transmitter generates a stream of very narrow pulses at periods of T and with amplitude unity (one), and then that this stream is multiplied by the audio signal. The resulting amplitude of each pulse is the same as it would have been under the sampling regime, but the multiplication allows us to introduce the notion of modulation, familiar to all who deal with radio or broadcasting transmitters.

The unmodulated pulse stream is periodic with period T, and it follows from elementary Fourier series theory that its spectrum is comblike with frequency components at multiples of 1/T. Now, if each component is regarded as a carrier, multiplication by the audio signal constitutes a modulation of each carrier by that signal, just as it would in a radio transmitter. The effect on the spectrum is the same as it would be in a transmitter, namely that each carrier becomes surrounded by sidebands reproducing the baseband spectral components in the audio waveform. These spectral relationships are shown in Figure 6.2.a.

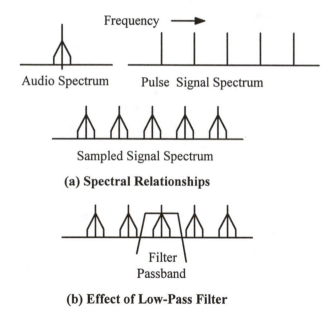

(a) Spectral Relationships

(b) Effect of Low-Pass Filter

Figure 6.2: Illustration of the Sampling Theorem

It should be evident that we could reproduce the baseband signal by passing the modulated pulse stream through an ideal low-pass filter as illustrated in Figure 6.2.b. If there is to be no interference from spectral components surrounding the harmonics of the sampling frequency, then clearly the relationship $1/T > 2B$ must hold, or $T < \frac{1}{2}B$, which is just the requirement expressed in the sampling theorem.

One should be careful to distinguish the spectrum of the sampled signal described from the spectrum we outlined in Chapter 2 for the digital transmission channel. Recall that the samples themselves are not transmitted. Rather the sample values are converted into binary numbers and the numbers are transmitted as a sequence of bytes, which is totally different from transmitting the samples themselves as voltage waveforms.

Now turning to quantizing noise, suppose each sample is to be transmitted as a number represented by four bits, which means that it would be quantified into 15 levels, say the voltages +/-7, +/-6, . ., +/-1, 0, to keep things simple. Thus if the actual signal level were say 4.8 V, this would be transmitted as 5 V, giving an error of 0.2 V. If a noise voltage of 0.2 V had been added to the original signal before quantization, it would have had the same effect as this error, and hence the concept of quantizing noise arises. Naturally, real systems quantize the samples to far more than 15 discrete levels and use far more than four bits per sample.

The quantizing noise level can be calculated by statistical techniques, but in practice this is fairly difficult. Uniform quantizing levels as described are rarely used. Fine steps are used at low signal levels and larger steps at higher levels, reflecting the probability that the signal spends more time at the lower levels.

Thus the more probable errors are kept small and the errors likely to occur less often are allowed to be larger. These characteristics are embodied in companding laws which have been standardized in many applications. Thus one will see references to A-law or C-law companding.

We can now look briefly at how these factors have controlled the bit rate necessary for high-quality audio transmission. It follows from the sampling theorem that to reproduce faithfully a signal with an audio bandwidth of 20 kHz, one must have a sampling frequency somewhat above 40 kHz, and 44.1 and 48 kHz are commonly found. For quantizing noise to be inaudible, it is generally held that the audio samples need to be quantified over greater than 50,000 levels requiring 16 bits per sample. Thus, typically, high-quality audio requires $2\times48\times16$ Kb/s or 1.5 Mb/s for a stereo channel. This is typical of the bit rates found in compact audio disk applications.

If we wished to broadcast this signal using one of the more robust modulation techniques described in Chapter 2, say quadrature phase shift keying (QPSK) at 2 bits per symbol, the bandwidth required would be about 400 kHz, which is much greater than the channel bandwidth required for an analog FM service and a substantial part of a TV channel. Hence, the need for bit rate compression is evident.

Psychoacoustic Masking

As mentioned, studies of human hearing indicate that not all the information in an audio signal is actually required for it to be perceived as high quality. Evidently some of the information is redundant and could be discarded before transmission. Just what can be discarded is determined by psychoacoustic models that mathematically model the processes of hearing and perception.

The studies used to develop the psychoacoustic models have revealed that many aspects of hearing can be explained by supposing that the ear decomposes the sound signals as if they were passed through a bank of band pass filters covering the audio spectrum. The pass bands of these hypothetical filters are known as the critical bands. They tend to be broader in bandwidth as their center frequencies increase. It is a characteristic of the critical bands that the perceived character of a stimulus remains relatively stable if its frequency components are confined to one critical band, but the character changes abruptly if it shifts into another critical band or if the spectrum spreads into adjacent bands. Thus the critical bands are relatively independent channels on which psychoacoustic models may operate.

The redundancy in hearing is manifested in the way that under certain conditions the signals in some critical bands may not be audible and therefore may not need to be reproduced. In particular, a high-level signal in one critical band may mask lower-level signals in adjacent bands if they fall below some threshold determined by the level of the dominant signal. The concept is illustrated by the diagram in Figure 6.3. This masking occurs while the louder signal is present, but also persists for brief intervals before and after the louder sound occurs. In addition, a dominant tone can mask noise in its own critical band.

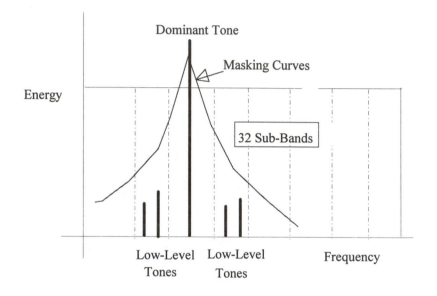

Figure 6.3: Principle of Psychoacoustic Masking

Broadly speaking, this masking effect is exploited in audio compression by sampling the signal level in each of the critical bands and dynamically assigning a bit rate to each according to the different resolution required by each over short periods. If a critical band contains a high-level tone, for example, it can mask quantizing noise in its own and adjacent critical channels. Hence, the quantizing noise in these bands can be relatively high. Additionally, if the adjacent bands contain signals low enough to be masked themselves, it may not be necessary to send data for them at all. The psychoacoustic model embedded in logic circuitry in the coder determines what level of quantizing noise is acceptable in each band. It adjusts the step size by which the sample is approximated by applying a scale factor to each band and then allocates a proportionate share of the total bit rate to the band.

The filter bank defining the critical bands does not normally exist in the coder. Rather, the coder operates mathematically on the spectrum found by taking a digital Fourier transform (DFT) of the incoming signal over a short period of time. The data applying to the signal over each short period are assembled into a package of data, known as a frame. In addition to the quantized sample values, each frame carries a digital code conveying to the receiver the bit allocation and the scale factor applying to each critical band.

The frame normally carries data for more than one channel, for example a stereo pair, but frequently there are up to six channels comprising surround sound channels or even separate language soundtracks. The sequence of frames comprises the packetized elementary stream for the audio service, as described in Chapter 4.

Practical Implementations
The most common implementations of this type of compression in broadcasting are Dolby AC-3 and a subset of the MPEG Standards. Dolby AC-3 has been adopted for the ATSC Standard in the United States. MPEG I is an early implementation for coding mono or two channel stereo services, while MPEG II has a multichannel capability. In addition, the MPEG specifications encompass three different levels. Level one

implements the simplest psychoacoustic model and provides a modest degree of compression, while Level 3 is much more complex and intended for applications where low bit rates are critical. Level 2 is intermediate between the two. In broadcasting, Level 2 is normally applied to program transmission, but Level 3 may be found in critical program distribution, such as the use of ISDN contribution circuits.

Dolby AC-3 analyzes each channel effectively over 256 audio samples, giving a time resolution of 5.3 mS (at 48 kHz sampling rate) over which the signal statistics are assumed constant. However, this resolution can be halved dynamically if transients are sensed by the coder circuitry. The filter bank has a uniform resolution of 93 Hz, from which the psychoacoustic model determines the critical band occupancy. On initialization the coder and decoder assume a set of core values for the model, and these are changed as necessary by the transmission of side information in the frame.

MPEG Level 2 operates over a block of 1152 samples giving a time resolution of 24 mS (at 48 kHz sampling rate). The processing emulates a set of 32 filters in the filter bank. Information on the scaling factors and bit allocations in each band is transmitted to the decoder in every frame.

Both standard systems are capable of multichannel operation on a variety of sampling rates. Both are capable of high degrees of compression. It is said that most people find the quality of a stereo service indistinguishable from the original at a bit rate of about 128 Kb/s. This is a reduction by more than a factor of 10 over the uncompressed bit rate.

Chapter 7

Video Compression: The Techniques Behind MPEG

The acronym MPEG stands for Moving Pictures Expert Group which is formally a part of a joint technical committee in the International Standards Organisation (ISO) and the International Electrotechnical Committee (IEC). MPEG has defined techniques for compressing the data rate required to transmit video services in digital systems. In broadcasting, these standards are used in the ATSC system in the United States and in the European DVB systems.

The initial specifications, commonly referred to as MPEG-1, had reasonably modest objectives, being intended for low-resolution pictures such as computer generated images transmitted or stored at data rates up to 1.5 Mb/s. MPEG-2, which essentially was completed only in 1994, extends the capacity to full broadcast standard video.

The key elements of MPEG address picture coding. MPEG-1 and MPEG-2 share many techniques in common to exploit the redundancies found in typical video sequences. Adjacent parts of a scene are very similar to each other, as are successive frames in the sequence of pictures that comprises the motion picture format. Where data to be transmitted change slowly, as with these characteristics, it is intuitively obvious that transmitting the changes is much more economic than transmitting everything and thereby repeating the static elements over and over. This is the common principle behind differential pulse code modulation (DPCM) systems. We shall show how these systems have been cleverly adapted by MPEG. Additionally, MPEG systems do not work on the picture elements (the pixels) themselves, but mathematically transform the pixels values using the discrete cosine transform or DCT. This transformation gives a stronger correlation between adjacent samples and many of the values generated can be discarded because their absence tends not to be noticed by the human visual system in the reconstructed images. The first property of the DCT makes the DPCM process much more effective, and the second substantially reduces the number of values requiring transmission.

Having defined picture encoding using these techniques, the MPEG standards next address the coding of the data for transmission. They provide for two common methods of data compression on the transmission link. These are run length coding and Huffman coding, the latter sometimes being described as entropy encoding.

All these processes comprise what is commonly called the "MPEG toolbox," a fitting name in that the standards allow a system designer to choose some or all of them and the extent to which they are applied in any application. The system is very flexible and effective. A few years ago, the standard transmission rate for broadcast quality video distribution was regarded as 144 Mb/s. Some users came to accept 34 Mb/s with the development of early compression systems, and now 8, 6 or even 5 Mb/s is generally considered adequate with MPEG-2. Let us now look at some of the elements of the standards.

Exploiting Differential Coding

Although MPEG does not use DPCM as such, the philosophy behind DPCM is very useful to develop an understanding of the MPEG techniques. Consider the simple DPCM system shown in Figure 7.1, and suppose it is transmitting some signal where the value of each successive symbol depends strongly on that preceding it, as happens for example with the pixels in a video signal. This system, as drawn, transmits only the changes between symbols, and the actual symbol values are recovered by the circuitry around the delay unit in the receiver, whereby the previously derived value is systematically added to the change value as it is received. The reader is invited to plug in a simple sequence of numbers to see that this rudimentary arrangement really does work.

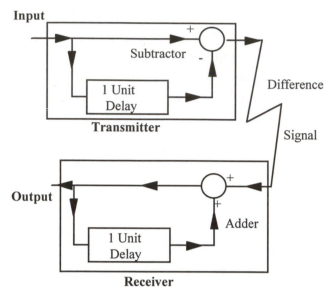

Figure 7.1: Rudimentary Differential Transmission System

In Figure 7.1, the function of the delay line in the transmitter is essentially to "predict" the value of the next symbol to be transmitted. If this prediction is reasonably accurate, then the DPCM system needs to transmit only the errors and the numerical range of the errors should be much smaller than that of the original symbols, leading to a considerable reduction in the data rate. The system as drawn uses only the value of the previous symbol as the prediction, but obviously if we could exploit some stronger property of the data stream to form a better estimate, the range of the errors and the data rate could be reduced even further. This is the essential principle of MPEG.

On this principle, we can replace the simple delay units in Figure 7.1 with "predictors" and arrive at the system shown in Figure 7.2.

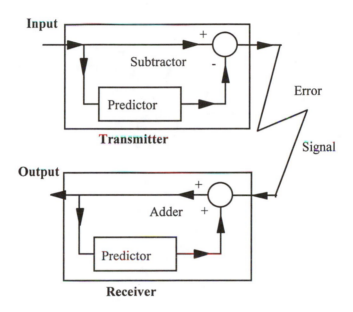

Figure 7.2: Elementary System with Prediction

Ideally, this system can work, but it suffers from the very serious deficiency that the predictors in the transmitter and receiver operate on essentially different signals. This makes designing the two predictors to track each other accurately very difficult, particularly when some processing of the transmitted signal, like quantizing, is introduced. Fortunately, there is a way around this by embedding the predictor in a feedback loop at the transmitter, as shown in Figure 7.3. This figure shows a quantizer in the transmission path to more accurately represent a real system and to illustrate the point that the two predictors here operate on the same signal.

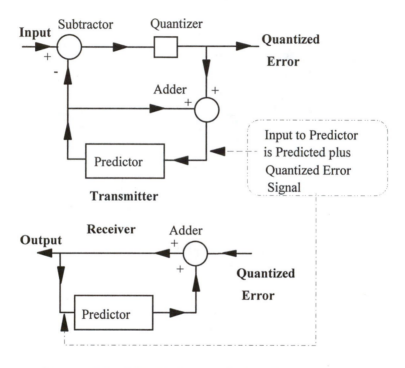

Figure 7.3: DPCM Transmission System

Figure 7.3 is the essence of the MPEG picture encoding system. The better we can make the predictors, the lower the data rate we have to transmit. The very clever part of the MPEG standards is in the prediction algorithms, but before addressing that, let us revert to the two coding methods mentioned, run length and Huffman coding, as they have a bearing on how the prediction algorithms have been chosen.

Run Length Coding

In some classes of data, particularly those arising from video pictures or facsimile scans, one often finds long strings of numbers with a constant value, especially zero. It is more efficient to transmit this as only two numbers, the constant value and the number of symbols in the run, rather than as the full sequence of the same number. This is known as *run length coding* or more often *variable run length coding* (VRLC). If one can transform the raw data in some fashion to make long strings occur often, then a reduction in the required transmission rate may well result. We shall see that MPEG achieves this by way of the DCT mentioned.

Huffman Coding

Huffman Coding, named after its inventor, is an example of entropy coding, and is simple in principle. Each symbol (or value) to be transmitted is assigned a binary code word of some number of bits. Words with a small number of bits are assigned to those symbols that occur more frequently, and longer words are assigned to symbols that occur only rarely. Thus, for example, fewer bits need be transmitted than in a system where all symbols are assigned words of equal length. A well-known example of the principle occurs in the Morse code. Huffman invented a systematic way of assigning words to values that has been adopted in the MPEG standards.

Huffman and run length coding are quite standard techniques in all classes of data transmission, and they are "lossless" in that, barring transmission errors, the original data can be regenerated exactly. This is not the case with the other techniques used in MPEG, in which the data (for example pixel brightness) may be reproduced only approximately by the predictors in the transmitter and receiver. The art of MPEG has been not only to make this "lossy" data reconstruction subjectively acceptable, but also to format the transmission parameters to make the lossless techniques particularly effective. The DCT which we now address, is an important element in this process.

The Discrete Cosine Transform

The discrete cosine transform is very closely associated with the Fourier transform, and like the latter, may be considered to extract the "frequency components" in a sequence of samples. In MPEG the DCT operates over blocks of eight rows by eight columns of pixels in the video picture, giving 64 numbers to be transformed in each block. The rows are runs of eight pixels taken from each of eight successive scanning lines and on the picture screen each block comprises a small square area of adjacent pixels. The process is illustrated in Figure 7.4, which purports to show a block of pixels transformed into a similar block of DCT coefficients. For simplicity, Figure 7.4 shows only 4 x 4 blocks. In a sense, the DCT coefficients across the block are a measure of the frequency components in a horizontal scan of the picture, and those down the block a measure of frequency components in a vertical scan. Higher-frequency components occur toward the right and bottom of the block and have large values only in busy scenes with a lot of detail. These values approach zero in scenes with little changing detail, such as pictures of the sky, for example. The overall average brightness of the pixel block is contained in the upper left-hand corner.

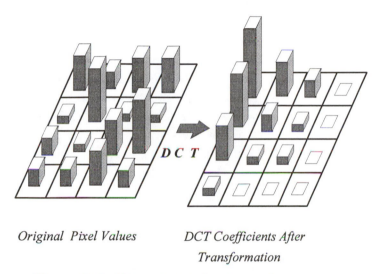

Original Pixel Values   *DCT Coefficients After*
                                 Transformation

Figure 7.4: Discrete Cosine Transform

Figure 7.4 attempts to show two common properties of the DCT that occur in most video applications. Normally the higher-frequency components diminish rapidly in value away from the top left corner, and in addition the correlation between values in adjacent cells is more marked than with the original pixel block. In viewing a reconstructed picture, the eye is increasingly less sensitive to the accuracy of the DCT coefficients toward the right and bottom of the DCT block, and these can be coarsely quantized or set to zero, according to the quality of reproduction desired. On the other hand, the coefficients toward the upper left corner govern the overall brightness of each reconstituted block of pixels and must be quantized finely and transmitted accurately. Otherwise, the brightness in the reconstructed scene tends to change abruptly from block to block leading to a checkerboard effect commonly called "blockiness."

To exploit these properties further, MPEG uses a scanning sequence in each block like that illustrated in Figure 7.5.b, rather than what might be considered the more natural sequence in Figure 7.5.a. Notice that this gives a smooth transition through adjacent values, avoiding large jumps in magnitude, and secondly, it can generate long runs of zeros or constant values where the higher-frequency DCT coefficients are coarsely quantized or set to zero. The first property favors efficiency in the DPCM process, as explained, and the second neatly fits in with the variable run length coding used on the transmission link, as mentioned.

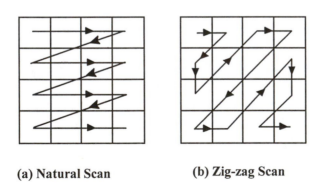

(a) Natural Scan   (b) Zig-zag Scan

Figure 7.5: The Zig-zag Scanning Sequence

The relationship between picture quality and the accuracy or transmission of the higher-order DCT coefficients adds another degree of flexibility in MPEG. It should be evident that in the inherent DPCM process, the prediction error will be greater in rapidly changing scenes and would ideally be transmitted at a higher data rate than otherwise. However, this is generally difficult in a world where most digital links have a fixed rate. With MPEG, the outgoing data are stored in a buffer that is emptied at the fixed rate but can be filled at a varying rate. If the buffer tends to overflow, a feedback loop into the picture coder causes the quantizing of the higher-order DCT coefficients to be made coarser, decreasing the volume of data and increasing the prevalence of long run lengths of fixed values and favoring the variable run length coding. Thus, normal picture sequences are encoded to the defined quality, but this quality declines for "busy" sequences or those with rapid motion. It is in just these circumstances, however, that the eye is least sensitive to picture quality. In a sense, MPEG achieves the desirable aim of matching its coding to its channel, of which the human visual system is part.

Motion Prediction

We have seen that the major essence of MPEG systems, arising from the principles of DPCM, is that they should transmit only the differences between a current block of data and a predicted block of data derived from what has gone before. The smaller this difference, the smaller the required data rate. In other than still pictures, the blocks of data will change position from frame to frame, and in predicting the values of a particular block, the MPEG predictor could very usefully use information on where the block came from in previous frames. Effectively, the predictor should keep track of moving blocks, as they change relatively slowly within themselves, compared to how quickly a block in a fixed position on the screen would change when motion is present. Think, for example, of the situation of a camera scanning a crowd scene. If we could track an individual face through the sequence of picture frames, the pixel blocks within that face might stay essentially unchanged in their values.

To give effect to this technique, MPEG systems transmit with each DCT block a motion vector indicating the velocity of the block across the screen in the horizontal and vertical directions. This tells the predictor in the receiver where the block will be next, or perhaps we could better describe this as the receiver knowing where the block came from to determine what its values were last time. There are various ways of estimating the motion vector, but most involve the transmitters moving the 8 X 8 block around a 16 X 16 "macroblock" in adjacent picture frames to find the position of best fit. This process, incidentally, is not defined by MPEG, which confines itself to requirements at the receiving end. Hence, the standards leave room for innovative solutions.

The MPEG Frame Structure

As described, the DPCM process embedded in MPEG has two problems. Essentially, it has to start somewhere—we cannot have a cascade of picture frames all derived from those that came before and not have a beginning. Second, transmission errors will propagate and affect all frames that come after the one in which the error occurs. In addition, in video, there must be frames capable of standing alone to permit editing, cutting, and splicing of program segments.

To satisfy these considerations, MPEG defines three types of frames to be transmitted. The first, called intra-frames or I frames, are stand-alone and independent of any previously transmitted frames. They are encoded entirely from within themselves and achieve only a moderate degree of compression. The second type comprises the predictive frames or P frames. These are coded using motion compensated prediction from previous I and P frames. They achieve considerably more compression than the I frames. Finally, there are the bidirectionally predicted or B frames. These are derived from past and future I and P frames and are not used for further predictions. They achieve the highest degree of compression.

The diagram in Figure 7.6 shows a typical sequence of frames and the interdependencies between them. Note that the frames are not transmitted in this order, as it considerably simplifies the receiver design if the P and B frames follow those on which they depend. The boxed material in Figure 7.6 shows the sequence in which the frames would actually be transmitted. (Note that this is a little simplified for clarity. In practice there would be B frames between P7 and I8 backwardly dependent on I8.) The out-of-sequence transmission does require frame storage in the receiver, and the standards do define a "simple profile" which omits the B frames.

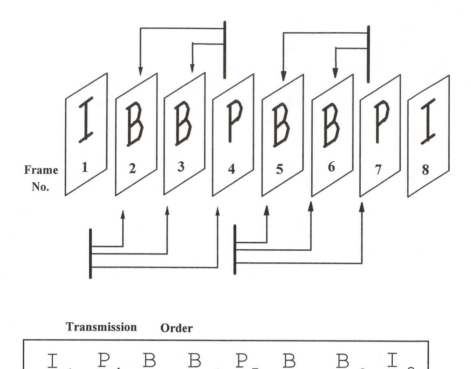

Figure 7.6: Frame Dependencies and Transmission Order

The MPEG standards define "profiles" and "levels." The profiles essentially describe the features of the toolbox that are included in a particular system, including some features not yet finalized such as scaling the fineness of quantizing according to the signal-to-noise ratio. The level largely defines the input source parameters, such as the maximum sampling rate. At present, readers are most likely to encounter the main profile, main level system suited to standard definition NTSC or PAL pictures.

Chapter 8

ATSC 8-VSB System

As mentioned in Chapter 2, the ATSC 8-VSB modulation system adopted in the United States for advanced digital TV is derived directly from linear modulation. Consider a baseband stream of digital symbols that has the form and spectral characteristics described in Chapter 2 as being necessary for band-limited digital transmission. If we were to apply direct linear double sideband modulation to this stream, the result would be a robust digital communication system. However, as in analog NTSC or PAL transmission, the bandwidth that this signal would occupy is twice that strictly necessary. As in the analog case, this arises because both the upper and lower sidebands carry the same information. Hence, one set of sidebands can be discarded without losing any information. In analog TV the use of vestigial sideband (VSB) modulation has been the standard method for many years. The ATSC standard now extends this to the digital domain.

In this chapter we look at some of the features of the ATSC system and show how they arise from the principles of digital transmission outlined in earlier chapters. The ATSC has adopted eight symbol levels for terrestrial transmission and sixteen levels for cable systems, where the environment is more benign in terms of interference and impulse noise. The discussion focuses mostly on the eight-level system as the eight-level and sixteen-level systems are very similar in principle.

Symbol Characteristics and Spectrum Shaping

It is instructive to consider the characteristics of an eight-level baseband system converted first to a double sideband (DSB) system by direct linear modulation, and then to consider what is required to convert this into a VSB system. Ultimately the system must occupy a standard 6 MHz TV channel, implying that the double sideband version can have a spectrum width of approximately12 MHz. The ideal rectangular spectrum would be like that shown in Figure 8.1.a, but remember from the discussion in Chapter 2 that we must have a symmetrical rolloff at the channel edges according to the Nyquist criterion introduced in that chapter. The Nyquist rolloffs are shown superimposed on the ideal in that figure, and we see that the ideal rectangular spectrum must occupy something less than 12 MHz, or equivalently, the ideal baseband channel filter must encompass a little less than 6 MHz. In fact, the ATSC has chosen a Nyquist rolloff of 0.31 MHz and an ideal filter bandwidth of 5.38 MHz. From Chapter 2 we see that this gives a symbol rate of 10.76 megasymbols per second. There is a compromise here, as a wider Nyquist rolloff would have given a sharper symbol less extended over time with less potential intersymbol interference, but it would also have lowered the achievable symbol rate. This choice keeps the symbol rate close to the maximum achievable.

88

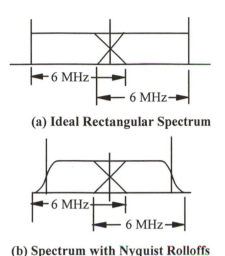

(a) Ideal Rectangular Spectrum

(b) Spectrum with Nyquist Rolloffs

Figure 8.1: Spectral Relationships in VSB

In converting this system to VSB, any vestigial slope characteristic that is symmetric about the carrier frequency would serve, even that shown in Figure 8.1.b, which one often sees in discussions of analog VSB. However, it makes sense to keep the filter characteristic symmetrical about the center of the transmission channel as practical filters that may be used to realize this characteristic in intermediate frequency stages all have that property. Accordingly, the VSB region has been defined as having the same form as the Nyquist roll-off at the top end of the channel. One sees that this puts the very bottom of the spectrum at 0.31 MHz below the carrier frequency, or in other words the carrier must be 0.31 MHz above the bottom edge of the channel.

Taking all this into account, we see that the 8-VSB channel characteristics are defined as in Figure 8.2. Actually, this figure shows the overall channel characteristic defined by the combination of filtering at the transmitter and at the receiver. We have shown that for an optimum noise immunity the filtering should be shared equally between the transmitter and receiver. Thus, for a transmitter or receiver alone the filter characteristic will have values that are actually the square root of those shown in the figure.

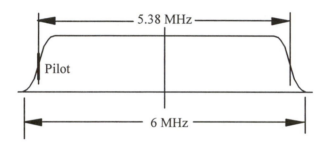

Figure 8.2: Practical VSB Channel Characteristic

Figure 8.2 shows the pilot signal at the carrier frequency, as required by the ATSC standards. The pilot is required as a frequency and phase reference for coherent demodulation at the receiver. Unlike their analog TV predecessors, digital VSB systems cannot be demodulated with an envelope detector and coherent demodulation is always necessary. The reason lies in the baseband symbol stream in which the symbol amplitudes can take on the normalized values +/-7, +/-5, +/-3, and +/-1 V. Note that these include negative values. By design, modulated analog TV signals are never allowed to go negative. A negative value implies a phase shift of 180 degrees in the carrier, and this cannot be detected by an envelope demodulator. Hence, coherent demodulation is required because of the baseband characteristics of the digital symbol stream.

It should now be apparent that the parameters defined for the VSB channel in the ATSC standards have their genesis in the simple principles of basic baseband digital transmission systems.

ATSC Framing Structure

The framing structure of the transmitted signal is an important aspect of the ATSC standard. As we shall see, the framing structure not only accommodates the transport stream requirements outlined in Chapter 4, but also plays an important role in mitigating RF propagation effects like multipath and impulse noise.

We mentioned in Chapter 4 that the basic transport stream packet for ATSC consisted of 188 bytes, including a synch byte. At the transmitter this is altered in two ways. First the synch byte is stripped off, leaving 187 bytes to be transmitted. Then 20 bytes are added to this for the Reed-Solomon error correction, giving 207 bytes to be transmitted in each packet, which at eight bits per byte amounts to 1656 bits. The trellis coding at rate two-thirds increases this to 2484 bits, or 828 symbols, since eight-level coding gives three bits per symbol. A special waveform, known as the data segment synch, is added to the head of this packet and occupies four normal symbol periods. Thus in total the modified transmission stream packet now occupies 832 symbol periods, or a total time of 77.3 usec at the symbol rate of 10.76 megasymbols per second. Each packet is now called a data segment.

Periodically, at intervals of 313 packets or 24.2 msec, a special data segment known as a field synch is inserted. The field synch carries specific data used by the adaptive equalizer in the receiver to estimate what echoes may be present because of multipath.

The form of the data segments is shown in Figure 8.3, which also shows the shape of the data segment synch waveform.

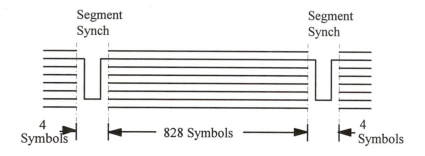

Figure 8.3: Form of Data Segments

Figure 8.4 shows the overall framing structure, illustrated by lining up the starting time of all the data segments over two frame synch periods.

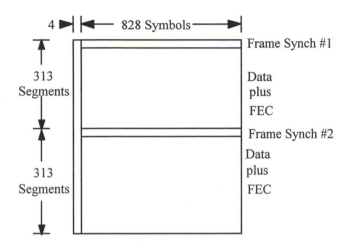

Figure 8.4: Overall Framing Structure

Securing Noise Immunity

The use of the data segment synch waveform adds greatly to the robustness of the system against impulse noise in particular. Recall that one of the worst aspects of impulse noise in digital transmission over RF channels is that it may cause the system to lose synchronization and significant time might elapse before the system regains that synchronization. The data segment synch waveforms are easily recognizable. They have only two widely spaced levels and are unlikely to be mistaken for data even during a high-noise event. The receiver locks its master clock to them using phase-locked loop techniques similar to those described in Chapter 4 for synchronizing transport stream packages.

The net effect of all this is that although the receiver may suffer data errors during a noise event it is much less likely to lose synch. Scattered data errors are normally corrected by the forward error correction provisions and might even have no effect on the perceived quality of the service, provided that the receiver retains synchronization.

Forward Error Correction

In addition to the immunity provided by the use of the data segment synchs the ATSC system employs two levels of forward error correction. The first is a Reed-Solomon (R-S) cyclic redundancy code, which, as mentioned, operates over the 187 data bytes and adds 20 bytes to each data segment. The second is trellis encoding at rate two-thirds. The R-S code can correct up to 10 byte errors per data segment. To ensure that a single burst on noise is unlikely to cause more than that number of errors in any segment, the bytes in a sequence of segments are spread out in time by being assigned more or less randomly across the segments in the sequence. At the receiver the bytes are reassigned to their correct segments, but it is unlikely that this reassignment will put together a run of incorrect bytes because they essentially come from scattered instances in time. This is the process of data shuffling mentioned in Chapter 3.

For each two bits on the input data stream the trellis coders generate three output bits, which conveniently constitute one symbol for the eight-level baseband coding. The system actually uses 12 coders so that the symbols produced by each are spaced 12 symbols apart in the symbol stream. Again, this adds to the immunity from impulse noise as it makes it less likely that a string of consecutive symbols presented to the decoder will contain a run of errors.

Adaptive Equalization

In Chapter 3 on the effects of RF propagation it was asserted that no digital broadcasting system could work satisfactorily if it did not have some built-in means of countering multipath propagation. The ATSC system applies a particularly robust method employing adaptive equalizers.

The general form of an adaptive equalizer is shown in Figure 8.5.

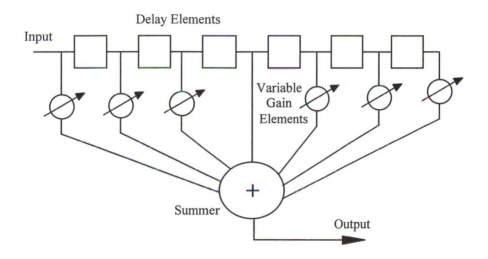

Figure 8.5: General Form of Adaptive Equalizer

It consists of a series of storage elements that hold the values taken at the receiver sampling instances over a long sequence of symbols from the stream being received. The samples move forward by one storage element as each new sample is taken at the input. The main output signal is usually taken from somewhere within the chain of storage elements. Samples from the storage elements before and after the main output element can be added to the output signals at levels that may be varied electronically. Varying the levels is referred to as changing the equalizer tap values, probably because some value is "tapped off" each contributing storage element.

The net effect of the equalizer is to add leading and trailing echoes to the main signal. Hence, if these echoes can be made the negative or inverse of the echoes arising from multipath, the detrimental effects of the multipath can effectively be canceled. To do this however the equalizer must be able to determine just what multipath echoes are present. It employs a "training sequence" contained within the frame synch segment that occurs after every 312 data segments or every 24.2 msec as described.

The training sequence is a burst of pseudo-random data occupying 511 symbol periods. The term *pseudo-random* indicates that it is affected by system defects just as truly random data would be, because it contains all possible level transitions, for example. In fact, however, its sequence of values is known precisely and stored in the receiver. Thus, every 24.2 msec the receiver can subtract this stored sequence from the values currently held in the storage element chain in the equalizer. The differences reveal the magnitudes of the echoes due to multipath. From this, the necessary tap values in the equalizer can be determined to negate these echoes, and of course these settings are recalculated every 24.2 msec.

The length of the training sequence is approximately 48 usec, and in principle echoes with this delay can be compensated. This corresponds to a path length difference of nearly 15 km or 10 miles. In practice echoes with magnitudes of up to 80% of the main signal amplitude have been successfully cancelled. That is quite an impressive performance.

Chapter 9

COFDM: The New Modulation

One of the techniques that best illustrates the advantages of digital over analog transmission is Coded Orthogonal Frequency Division Multiplexing (COFDM). The new method promises great improvements in the applications of radio propagation by doing away with multipath effects, allowing on-channel repeaters, and permitting a new service area configuration of transmitters known as the single frequency network. To illustrate its advantages consider some of the real life problems that arise in typical broadcasting situations.

For example, consider setting up links for a TV outside broadcast coverage of the running of a marathon. One could try hiring space for microwave repeaters on city buildings or using helicopters. Alternatively, COFDM would permit using reflecting signals from those buildings to create a transmission path back to the studio.

Or consider covering a large city for TV or radio broadcasting. The present practice is to seek out a very high site and install a tall tower and powerful transmitter. Alternatively, with COFDM it is possible to scatter a few small transmitters across the city, all operating on the same frequency and carrying the same program, and giving almost perfect coverage.

Or in delivering services from a satellite, the signals are often blocked into some streets by high buildings. A solution might be to put satellite dishes on some roofs, and distribute the received signals by cable. If COFDM were used, however, the blocked sites could be served by a small repeater, retransmitting the satellite signal on its own frequency into the shadowed areas.

COFDM is used for terrestrial radio and TV broadcasting in Europe and some other countries, but not as yet in the United States. In the next few years we can expect to see it used extensively in outside broadcasting applications.

Principle of COFDM

COFDM is essentially a new method of transmitting high data rate digital signals over radio channels, or for that matter over wire and cable links. To make the numbers specific, let us look at how COFDM might be used to transmit a digital stream at a data rate of say 1 megabit per second (1 Mb/s).

The principle of COFDM is to break this single bit-stream into a number of parallel bit-streams and to transmit all of these lower bit-rate streams on a very large number of carriers side by side within the one channel. For the 1 Mb/s stream in this example, we may choose to break it up into, say, 500 parallel streams each carrying a bit rate of 2 kilobits per second (2 kb/s). Each of those

2 kb/s second streams is then modulated on to one of 500 carriers spaced 2 kHz apart. With the 500 carriers spaced at 2 kHz, the total channel occupancy in this example would be 1 MHz.

The numbers applying to the carrier spacing and the bit rate on each are chosen according to the principle of *orthogonality*. Orthogonality is a mathematical concept, but it simply means that the arrangement of the carriers is such is that each can be demodulated without interference from any of the others. To provide for this, the digital symbols modulating each of the carriers are shaped so that any carrier has a spectrum of sidebands around it very similar to the form shown in Figure 9.1.a. The spectrum shows a central peak at the carrier frequency and points of zero magnitude at sideband frequencies corresponding to multiples of the symbol rate. Now if we space a number of these carriers apart so that the peak of any one falls on the spectrum zeros of all the others the overall frequency spectrum appears as shown in Figure 9.1.b.

It may be apparent that any carrier in the ensemble of Figure 9.1.b can be demodulated without interference from the others. Since the peak of any carrier coincides with the spectral zeros of all the others, the peak value is unaffected by what amplitude the others may have from moment to moment. This is the principle of orthogonality. The reader will no doubt observe that it is very similar to the method described in Chapter 2 for avoiding intersymbol interference in baseband systems. Here, we are working in the frequency domain, whereas previously the techniques applied in the time domain.

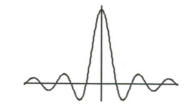

(a) Spectrum Around One Carrier

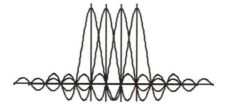

(b) Spectrum Around Several Carriers

Figure 9.1: Orthogonal Carrier Spectra

Most of the modulation schemes found in single-carrier systems are used without changing the essential principle. Suppose that, in our example, we had chosen to use 16 QAM on each carrier. With 16 states available, the data capacity is 4 bits per symbol. For our 1 Mb/s example, therefore, we need to transmit 250,000 symbols per second. If we choose to use, say 1000 carriers, the modulation rate on each must be 250 Hz (strictly, 250 symbols per second), and to preserve orthogonality, the carrier spacing would be 250 Hz. Thus the total bandwidth required is now 250 kHz. There is thus a simple relationship between the bit rate, the bits per symbol, the symbol rate, the carrier spacing, and the required channel bandwidth.

At this stage many readers will rightly regard COFDM as quite complex, and it is that complexity that has delayed its commercial application for over two decades. Soon, we will show how modern digital signal processing has come to the rescue, but first it will be worthwhile outlining the advantages of COFDM that have made tackling the complexity worthwhile.

Properties of COFDM

One of the outstanding characteristics of COFDM is its resistance to multipath propagation. First, let us look at this from the point of view of the frequency domain. A frequency sweep applied to a channel showing multipath effects would normally look a little like Figure 9.2 which shows the path loss plotted against frequency. There is normally a dominant echo signal on such paths which, in the frequency domain runs into and out of phase with the main signal. Consequently, the frequency sweep shows broad peaks where the echo is in phase and narrow but relatively deep minima where the echo is out of phase.

With analog modulation, these dips can cause signal distortion by removing a significant part of the energy in the sidebands. More particularly, if they happen to attenuate the carrier, they may greatly disrupt the process of demodulation, with results that listeners to broadcast FM channels in cars have often experienced.

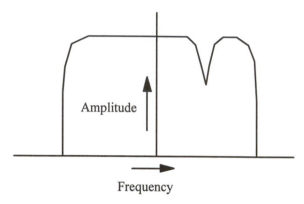

Figure 9.2: Channel Frequency Response with Multipath

Studies of multipath propagation in urban and suburban situations have shown that the frequency ranges over which the dips in the frequency response are considered deep span from a few hundred hertz to a small number of kilohertz. If one could arrange a transmission system so that energy taken out of the signal over these ranges always caused little distortion, then the effects of multipath would largely be overcome.

The COFDM carrier spacing and bandwidth can be arranged so that the dips in the frequency response of typical propagation paths severely attenuate, or may take out completely, only a few of the possibly many hundreds of carriers that may be being transmitted in the channel. Since the bit-stream is spread across all the carriers, some carriers going missing means that only a few bits are absent or are wrongly demodulated in the delivered bit-stream. Furthermore, it is common to distribute the bits in a pseudo-random manner across the carriers. Thus, a missing carrier gives rise to an essentially random, rather than a periodic, pattern of missing bits. A reasonably robust forward error correction scheme can accurately correct or replace the missing bits in most circumstances. This scattering of the bit stream across carriers accounts for the term "coded" in the title.

Now let us look at the effects of multipath in the time domain. In urban and suburban environments the echo signals in most propagation paths are found to arrive within a few tens of microseconds of the main signal. Hence, if we could arrange the symbol period of the digital signals on any carrier to be much longer than this, the echoes from earlier symbols would have died out within a few percent of the duration of any particular symbol. This argues for long symbol durations and slow consequent signaling rates, which is of course just the situation that applies to each carrier in the COFDM channel. In fact, all applications of COFDM apply a "guard band" that is a short time over which the receiver ignores the demodulator output at the beginning of each symbol period, allowing any path echoes from previous symbols to die out before the receiver tries to estimate the signal values. Typically, guard intervals might encompass 10% to 25% of the symbol duration.

After the guard interval, of course, the spurious multipath signals cause a symbol to suffer from interference from delayed echoes of itself. However, it is easy to visualize the effects of this as causing a constant error in phase or amplitude, which once the transients have died out in the guard band is constant from symbol to symbol. Hence, all of the carrier modulation schemes used with COFDM represent symbols by defined changes in amplitude and phase, rather than by using specific fixed values. To a first order approximation, the starting and finishing points are affected equally by interference to themselves outside the guard interval,

105

and hence the change between the two is largely unaffected by the interference. Since the change defines the symbol, the probability of identifying the correct symbol remains high. Transmitting the changes like this is known as differential code transmission.

Figure 9.3 illustrates the concept. Here, suppose that a carrier phase shift of 90 degrees between one sampling instance and the next represents the bit pair 01. This may happen, for instance, with differential quadrature phase shift keying (QPSK) modulation. In that case a phase shift of 180 degrees might represent 10 and so on. The two large vectors in Figure 9.3 represent the carrier phases before and after the transmission of the particular bit pair. To these we have added two small vectors representing echoes of the main vectors due to multipath. Notice that, at least to a first order, the echoes do not change the overall phase shift of the carrier between the two sampling instances.

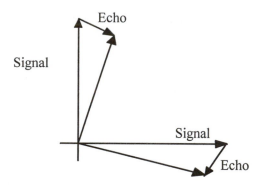

Figure 9.3: Effect of Echo on Differential Phase

Now let us expand on this for echoes from moving reflectors. In this case the path difference between the main signals and the echoes will vary over the time between the sampling instances, and the phases between the signal and echo at the two instances shown in Figure 9.3 will be different. In this case the echoes do have a first order effect on the differential phase shift, and will act to reduce the noise immunity of the system.

Notice that the change of phase between the main and echo signals over the symbol period constitutes a change in frequency between the two, caused by the fact that the reflector is moving. This is the well-known Doppler shift due to reflections off a moving target. It is also the cause of aircraft flutter in analog TV systems, whereby the echo's running in and out of phase with the main signal causes fluctuations in picture brightness.

Doppler shift is important when COFDM is used for mobile applications, such as Digital Audio Broadcasting (DAB). As we shall see, the DAB standards adopt different operating modes based on the expected incidence of Doppler shifts in different circumstances.

In summary, we see that two very effective techniques combine naturally in COFDM. The use of a great many carriers means that a few missing is readily accommodated. Second, the use of long symbols ensures that transient multipath effects disturb only a small part or their time slots.

It is worth noting that a somewhat incidental, but still very useful, property arises from the multicarrier nature of COFDM. Suppose that a narrow band system is interfering with a COFDM channel. It is easy to arrange matters so that the transmitter suppresses the affected carriers, by setting their amplitudes to zero, and the receiver ignores anything found on their frequencies. Conversely, if a COFDM signal interferes with another service, the offending carriers can be dropped in a like manner. In digital TV applications, for example, removing those COFDM carriers which correspond to the vision, color, and sound subcarriers of the analog TV system has been found most effective in permitting co-channel operation of digital and analog services.

Single Frequency Networks

The capability of forming Single Frequency Networks (SFNs) is one of the most intriguing aspects of COFDM. To see how these networks operate consider a service area surrounded by several transmitters so that a receiver in the area can "see" or receive from some or all of the transmitters. Let us suppose that the transmitters are carrying COFDM, are all operating on the same frequency, and are all carrying the same digital data stream. With conventional modulation systems this would produce intolerable interference. But consider what happens with COFDM.

The receiver sees the strongest signal as the main signal. The other signals on the same frequency and with the same data arrive at slightly different times according to the propagation distance from each transmitter. Thus the receiver overall sees a main signal and several identical smaller versions of it slightly scattered in their times of arrival. But this is just the situation that prevails under multipath conditions from a single transmitter. Hence, all the COFDM properties for defeating multipath effects are called into play as described, and the receiver output is essentially perfect and error free, provided only that the scatter of arrival times remains within the guard band time interval. This is easy to arrange in the design of most networks.

The obvious advantage of SFNs is that we can use three or four geographically spaced transmitters to cover any particular region. Thus, locations that may be shadowed from one will almost certainly be covered by the others. The service area thus becomes very solidly covered with almost no shadow areas from buildings, hills, or trees. However should there be any small location that somehow is not filled, one can feed it with a "gap filler." A gap filler is a simple non-frequency-changing repeater and could consist merely of an amplifier connected between receiving and transmitting antennas with enough isolation. It amplifies and retransmits the off-air signals and, as far as the receiver is concerned, is just another transmitter in the SFN.

In a SFN the carrier phases and data timings at each transmitter must be locked to reference sources, of course. These references are generally available where modern telecommunications links like ISDN are used for signal distribution to the transmitters, but interestingly, the GPS satellite system has also been employed as a timing reference.

Generation of COFDM

Conceptually, the simplest way of generating the COFDM signal would be to use a set of parallel oscillators and mixers as illustrated in Figure 9.4.

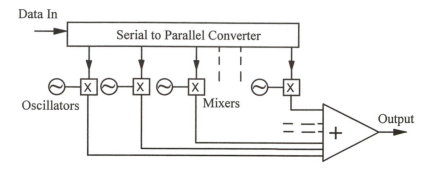

Figure 9.4: COFDM Generation in Principle

Given that there may be hundreds or even thousands of carriers involved in any system, this method is obviously impossible in practice. In fact, although the advantages of COFDM were recognized as far back as 1971, its commercial development did not commence until around 1990. The breakthrough came in the form of powerful digital signal processing with fast integrated circuit chips. The solution is both simple and elegant.

If one writes out the mathematical expression for a group of carriers modulated as in COFDM, it looks remarkably like the expression for Fourier transforms found in most textbooks. In fact one can show that it is equivalent to modulating a single carrier with a signal given by a Fourier transform taken across the data streams on all of the carriers. This is just the process to which high speed digital signal processors are now being applied, and it can be made very rapid by using algorithms like the fast Fourier transform. Hence, the generation of the COFDM signal can now be realized as easily as shown in Figure 9.5. (Strictly, the process involves the inverse Fourier transform, but this is just a mathematical quibble.) The receiver, of course, simply uses the inverse digital transformation. The processor speeds now becoming available mean that this technique can be applied to real time applications like those outlined in the next section.

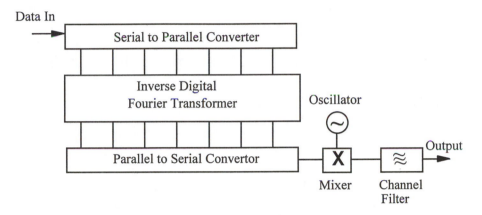

Figure 9.5: COFDM Generation in Practice

Current Applications

Currently, the most developed applications of COFDM lie in the field of broadcasting for sound, (DAB), and for Digital Terrestrial TV Broadcasting (DTTB). International technical standards have recently been agreed for both.

For DAB, commercially operating systems were commissioned in London and Stockholm in 1996. Large experimental systems exist in Canada, Germany, and other European countries. All use the SFN concept and have typically some four to six transmitters.

The DAB standards permit several transmission modes, but typically they operate with several hundred carriers in a 1.5 MHz channel with a bit rate of approximately 2 Mb/s. With MPEG or AC-3 audio coding, each system can carry from five to fifty separate audio programs with the best quality equal to that available from compact discs.

The initial DTTB applications employed 2000 carriers in an 8 MHz European TV Channel. Recant developments in digital signal processing chips have allowed the number of carriers to be increased to 8000. As this gives a longer symbol period it leads to easier design of SFNs. The systems invariably use MPEG coding and can carry up to five or six simultaneous programs.

Extensive experiments have been carried out in Japan on TV outside broadcasting applications, and equipment is becoming available commercially. Finally, it may be worth noting that the modulation systems used for experimental systems for pay-TV over phone lines also employ a variant of COFDM.

Chapter 10

Satellite Systems

The area in which digital transmission seems set to have its greatest impact is that of direct-to-home (DTH) television broadcasting from satellites. This field has been restricted in the past by the high cost of transponder capacity and the fact that analog technology permitted the transmission of only one program in each transponder. With digital transmission, as we shall see, up to five or six programs can be carried in each transponder. In addition, the cost of powerful satellites in orbit has fallen dramatically. Hence, we are now seeing subscription DTH systems starting up in the United States and Europe and making a huge choice of programs available to their customers. This chapter describes the techniques and design parameters underlying these systems.

System Characteristics

Figure 10.1 shows the overall block diagram of a satellite delivery system. An important point to note about it is that it shares many characteristics with the terrestrial systems we have discussed in previous chapters. Thus the video and audio coding standards and equipment, the formulation of the transport streams, and the forward error correction schemes are all common. Only the actual delivery system is different.

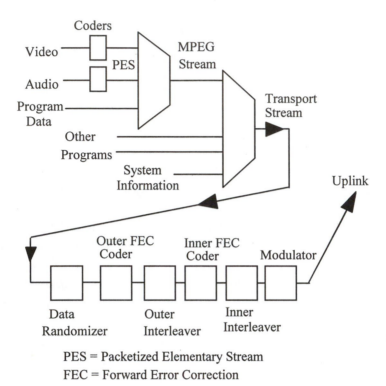

PES = Packetized Elementary Stream
FEC = Forward Error Correction

Figure 10.1: Typical Satellite Configuration

The advantage of all this commonality, of course, is that domestic receivers for the two systems contain many identical components and systems. To change from one method of delivery to another requires little more than a switch from one RF front end and antenna to another. It also means that satellites can very easily be used for program delivery to terrestrial rebroadcast transmitters or cable head-ends with little or no need to decompose the transport streams and bit encoding. As these common elements are all discussed elsewhere in the book, this chapter will focus on the satellite system itself as a digital transmission medium.

Satellite Parameters

The frequency bands used for DTH satellite services have traditionally been based around 3 GHz, known as C-Band, and 12 to 14 GHz, known as Ku-Band. The 3 GHz band has the advantage that the signal is only marginally attenuated by rain, and it has become the preferred option for delivery to rebroadcast transmitters or head ends. The low frequency and long wavelength dictate that relatively large receiving antennas must be used. These typically range in size from about 3 to 7 meters (10 to 25 ft), which is not a major disadvantage in these applications. However, these antennas are far too large for this band to be used for general DTH purposes.

The 12 GHz band does suffer from rain attenuation, but with modern high-powered satellites can provide satisfactory performance in most regions of the world with receiving antenna diameters down to about 600 mm or 2 ft. This is suitable for DTH services, and accordingly the 12 GHz band has become a valuable public resource.

The impending value of the 12 GHz band for broadcasting was foreseen relatively early and in 1977 it was apportioned to all the nations of the earth by means of an international treaty. The literature on satellite broadcasting often makes mention of this famous conference, referred to universally as WARC'77. WARC stands for World Administrative Radio Conference. A series of these conferences was held under the auspices of the International Telecommunications Union (ITU), an arm of the United Nations in Geneva.

WARC'77 allocated not only frequencies in the 12 GHz band to each nation, but also orbital slots. An orbital slot is the position a satellite may occupy in the geostationary orbit above the equator. Satellites in the geostationary orbit are at an altitude of about 32,000 km or 20,000 miles above the earth. At this altitude the orbital period is 24 hours, so that as the earth turns the satellite goes around the earth at the same angular rate. Hence, the satellite appears to stay above some fixed line of longitude. This line of longitude denotes the orbital slot allocation. The WARC'77 allocations have been revised to take into account recent developments such as digital technology, but the principle remains the same.

Geostationary satellites appear to remain in a fixed position in the sky, and hence remain within the beamwidth of high gain receiving antennas. Some drift in orbit does occur but is corrected as necessary by on-board thrusters controlled by the satellite operators from the ground.

Transponder Characteristics
In principle the function that the satellite has to perform is disarmingly simple. It needs only to receive the uplink signal, change its frequency, amplify it, and retransmit it back toward the earth. The equipment that does the frequency translation and

amplification is collectively known as a transponder. In practice, of course, the characteristics of transponders are far from simple.

The greatest contributor to complexity in operating transponders is the necessity to squeeze out of them the absolute maximum of transmitter power. Hence, they are conventionally run at power levels where, as amplifiers, they are approaching saturation and consequently exhibiting quite nonlinear transfer characteristics. As such they generate spurious radiation into the adjacent channels served by other transponders on the satellite. The transmitted power limit is nearly always determined by the need to keep this spurious radiation below some acceptable level. The operating power level of the transponder is often described in terms of its "output backoff," quoted as so many decibels. Thus a 3 dB output backoff indicates that the transponder is operating with an output power 3 dB below its saturated level.

The nonlinearity of transponders has generally meant that each could carry only one channel, for if two or more carriers were present the level of intermodulaton products became intolerable at relatively low power levels. Output backoffs of 12 dB or more can be required in these circumstances. Second, analog transmissions invariably used FM modulation as it is very tolerant of nonlinear effects such as amplitude compression in the transponder. Consequently, the bandwidth required to carry a single channel is relatively enormous, being typically 25 to 30 MHz. FM has a high noise immunity, so that effectively its use represented a trade-off of bandwidth for power. Even the best compromise, however, was not particularly good and it seems certain that DTH services would never have been really economic in an analog environment, although some did go into service. Enter digital, however, and the scene has changed enormously.

Digital transmissions through satellites use QPSK modulation as described in Chapter 2. Although we said there that the name is not strictly accurate, the method is close to a pure phase modulation, which implies that the carrier amplitude is relatively constant. A glance at the constellation diagrams in Figure 2.9 shows that the carrier has the same amplitude for all of the constellation positions, whereas this is not true of modulation schemes giving more bits per symbol, such as 16 QAM, for example. Hence, QPSK is by far the method most resistant to the effects of amplitude compression in nonlinear transponders and finds almost universal application in satellite DTH systems.

QPSK provides a transmission capacity of two bits per symbol. The transponder bandwidth on modern satellites is normally in the range of 26 to 72 MHz, with DTH applications tending to favor around 33 MHz. After allowing for band edge effects and the Nyquist frequency rolloff discussed in Chapter 2, one finds that the transmission rate through a typical transponder is about 26 megasymbols per second. The raw transmission rate is thus 52 Mb/s. The FEC rates for forward error correction can be chosen in the range rate $^1/_2$ to rate $^7/_8$ so that the payload bit rate lies in the range of about 26 to 42 Mb/s. Given that with MPEG coding a good-quality TV picture requires about 6 Mb/s, we see that under a digital regime, a satellite transponder can carry from four to six programs depending on the circumstances. This has totally changed the economics of broadcasting from satellites.

Sound Broadcasting

What the favored technology for sound broadcasting may be in the future is far from clear. There are two major competing systems. One is an adaptation of the Eureka 147 DAB standard described in Chapter 12. This uses COFDM modulation that has a far from constant amplitude in the transmission channel. In fact it exhibits a peak-to-average ratio of around 10 dB. Accordingly, it is very sensitive to amplitude compression in transponders, and these must be run at a considerable output backoff in the satellite. Additionally, a typical ensemble occupies a bandwidth of only about 1.5 MHz so that it makes very inefficient use of the transponders on the current generation of communications satellites. If Eureka 147 is to be broadcast via satellite in the future it will most likely be from purpose-built satellites.

The great advantage of using COFDM would be the capability of using repeaters on the ground to fill in points or regions that may be shielded from the satellite signal by mountains or tall buildings for example. The chapter on COFDM outlines the concept of single frequency networks, which is readily extended to a satellite service. Thus active repeaters on the ground might receive the satellite signal, amplify it, and rebroadcast it on the same frequency into the shielded area. It would be possible even to radiate the signal into road tunnels and the interiors of buildings by this means.

Despite the problems with transponder nonlinearity, Eureka 147 has been transmitted successfully through communications satellites in some experimental trials, notably in Australia and Brazil. Satellite broadcasting has some obvious attractions for large sparsely settled countries such as these.

119

The alternative method of sound broadcasting via satellite is being put into service by the Worldspace organization. The aim of Worldspace is to provide an alternative to the international HF radio "short-wave" systems used by such organizations as the Voice of America and the BBC Overseas Service. Hence, one of its imperatives is to be able to serve cheap handheld receivers of a type that would be affordable in the developing countries where these broadcasters find an important part of their audience.

Worldspace will use three satellites operating in L-Band, at around 1.5 GHz, and with the orbital slots spaced evenly around the equator. The audio coding is MPEG layer 3, to obtain very high degrees of compression such that normal program material requires a bit rate of only 11 kb/s. The modulation is QPSK with the bit rate kept deliberately low to make the necessary bandwidth in the receivers small. The RF noise level in the receivers is proportional to the bandwidth, and thus receivers with narrower bandwidths can operate at lower received carrier levels.

The first satellite in the system, Afristar, was successfully launched in October 1998. Short wave broadcasting is the oldest technology used in the industry. It is interesting that it will probably be the first to be displaced by a combination of the newest technologies in digital transmission and satellites.

Chapter 11

Engineering the Transmission Channel

There are similarities and also some significant differences between the new digital and the old analog requirements in the engineering principles that apply to high-powered transmission systems. Some of the aspects that used to be considered vitally important, such as the very strict requirements on impedance matching in antennas, are now less so. Alternatively, digital transmission brings with it other factors, such as peak voltage clipping in transmitter power amplifiers that were barely considered in the analog world. This chapter is essentially a survey of the differences between the two systems and is intended as a guide for those involved in the design and maintenance of the new high-powered digital systems. The chapter focuses on television, since it is in this area that one finds the subtle differences that those skilled in analog technologies will most usefully come to appreciate. There is very little similarity between digital and analog sound broadcasting technology, at least as world standards now stand. It should be said, however, that the transmission requirements for all digital services are much the same, and much of what is said herein for television also applies to digital sound broadcasting.

Power Requirements

How do digital transmitter powers compare with analog in the same service area? As outlined in the chapter on radio frequency propagation, the answer depends somewhat on the percentages of locations and of time that the system designer may choose to be covered to give a satisfactory result. The maximum ERP permitted will often be mandated by regulatory authorities based on considerations of both coverage and interference to other new or existing services. Thus in the United States, the FCC has determined a table of equivalent radiated power (ERP) values for all the new digital services associated with the current NTSC transmissions.

The discussion of comparative power requirements must take into account the different characteristics of the transmitted waveforms in the digital and analog systems. In NTSC and PAL TV, the peak voltage during the horizontal synch pulses is well defined and constant, whereas the voltage levels during the active picture intervals are not, as they depend on the scene being televised. Hence, the peak synch power has become the accepted way of describing the transmitter power or ERP for these systems. By contrast, in the digital systems the average value of the transmitted signal is well defined but the peaks can be described only statistically, and as we shall see, rare peaks may even be clipped off in the transmitter power amplifiers. Hence, digital transmissions are described in terms of their average power rather than in terms of peak power. When we compare powers for the two, therefore, we are generally comparing digital average power with analog peak synch power.

With what has been said about coverage probabilities and the actions of regulatory bodies, it is probably unwise to generalize about the differential power requirements between digital and analog transmissions into the same service. However, as a rough rule of thumb indicating what might generally be expected, one can say that the digital average ERP will generally be some 10 to 12 dB below the analog peak synch ERP in the same band. This also applies to the transmitter powers if they run into the same or an equivalent antenna system.

Now with the digital services, the transmitter power amplifiers have to be able to handle all but the very rare peak signal levels that may occur. We shall discuss the statistics of these rare peaks later, but for the ATSC 8-VSB system this implies a transmitter power capability of about 7 dB above the average, and for COFDM systems some 10 dB above the average. Hence, overall, the transmitter peak power capability required will lie between being one-half and equal to that of an analog transmitter providing the same coverage. As power supplies, line sizes, and other critical elements in a transmission system are broadly determined by the peak transmitter power, one can expect the overall size and complexity of a digital transmission system to be rather similar to the analog one it might replace.

There is one major exception to this rule, where, as we saw in Chapter 9 on COFDM, single frequency networks might be adopted. These would replace one single high-powered transmitting station with a network of low-powered transmitters all on the same frequency. This, however, does not seem to be the trend in the initial TV applications, although it is occurring for DAB in Europe. Single frequency networks are readily realized for

123

COFDM systems, but are yet only at the experimental stage for 8-VSB.

Peak Voltage Levels

Let us look at the question of peak-to-average voltage levels more closely. Recall that in Chapter 2 on the basics of digital transmission systems, we modeled a typical transmitter by passing the data signal in the form of voltage impulses into a filter that defined the symbol shape and channel spectral characteristics. The amplitude of the voltage impulses reflected the particular binary value that applied to each. With real data these amplitudes are therefore essentially random in that they do not follow any particular pattern. Hence, we have here a random signal exciting a narrow-band filter, and the output can be expected to have some of the character of random noise. Recall also that, except at the receiver sampling instances, the symbol waveforms were allowed to interfere with each other to a considerable extent. Hence, the voltages at any time, other than at a sampling instant, would be the sum over very many symbols leading and trailing in time. It is likely then, that very high peak voltages could occur if some data sequence caused all the voltages from these symbols to be in phase or nearly so. As the data sequences involved would necessarily be long for high peaks, these peaks should be rare, occurring less frequently as the voltage level increases.

For 8-VSB the voltage impulses in our model can take on any of eight levels. The analogy between the model and COFDM is not so obvious, but in Chapter 9 on COFDM, we saw that it could be generated by multilevel modulation from the output of a Fourier transformer operating on the input data. Thus, COFDM can also be related to our simple model. In TV the impulse levels can take up to 8000 values in this case. Accordingly, we might expect the transmitted COFDM signal to be even more noise like than 8-VSB. Indeed, this is so. Figure11.1 shows the probability of some level being attained for both 8-VSB and COFDM. The probability is plotted against the level expressed in decibels above the average. These are measured results taken on real systems. On the basis of these measurements the clipping level in transmitters is usually set at between 6 and 7 dB above average for 8-VSB and between 9 and 10 dB for COFDM.

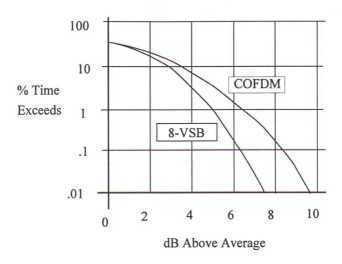

Figure 11.1 Peak-to-Average Probabilities

It is certain that from time to time the input signal will exceed these levels and clipping will occur, although this will be a rare event. When clipping occurs it will most likely cause an error or a short burst of errors. However, as we have seen, all practical digital transmission systems include powerful forward error

125

correction schemes, and hence these rare instances have almost no practical effect as far as the overall quality of service is concerned. This would not be true, of course, if clipping events occurred too frequently. The levels quoted are an accepted compromise between reducing the robustness of the transmissions on the one hand and having to size the transmitters, feeders, and antennas for very high voltages on the other.

There has been some expectation that the peak voltage pulses in digital systems, being brief as well as rare, may not be maintained long enough to cause voltage breakdowns in feeders and antennas. This may have allowed some relaxation in the voltage rating of these components in digital working and permitted them to be run at higher peak powers than would otherwise have been the case. It is found from classical physics, however, that the time constants for voltage breakdowns in air are of the order of 200 nanoseconds. We know from the discussion of symbol shapes in Chapter 2 that the half amplitude duration of isolated pulses is inversely related to the bandwidth of the transmission channel and for DTV is around 100 nanoseconds. These two figures are uncomfortably close. The conclusion must be that it is unwise to relax the conventional peak power ratings for digital transmissions.

This argument may not be true for multichannel systems where several transmissions may be carried in the same feeders and antenna. Highly specialized statistical analyses suggest that some relaxation may be permissible, allowing the actual peak voltages to exceed the conventional transmission plant ratings by a few decibels. These studies are at a very early stage, however.

Transmitter Spurious Radiation

In the analog NTSC or PAL systems, the radiated signals contain a small number of dominant carriers, notably at vision, color subcarrier, and sound frequencies. In all digital systems, however, the spectrum is considerably more densely occupied. As would be expected therefore, there is more opportunity for signal mixing to take place because of minor nonlinearities in the transmitter power amplifiers. This is made worse by the process described above whereby the power amplifiers may be allowed to run into saturation at some defined clipping level on peaks. Hence the level of spurious radiation into adjacent channels is likely to be relatively higher in digital transmitters than in analog.

This problem is compounded by the frequency planning and channel allocation policies that will necessarily apply during the period when digital and analog services must coexist until the latter are eventually closed down. To secure enough channels for this period the regulatory authorities are assigning services into adjacent channels in the same service area. In a purely analog situation the adjacent channels were considered "taboo" because of the propensity of analog services to interfere with each other. With mixed services this adjacent channel working becomes permissible if the relative levels are chosen carefully. It is compromised, however, when the transmitters emit significant levels of spurious radiation into the channels adjacent to those to which they are operating. Accordingly, all regulatory authorities have assigned very severe limits on the permitted levels of spurious radiation from digital transmitters. The mask adopted by the FCC in the United States is shown in Figure 11.2. Hence, it is likely that all transmitters will have to be fitted with what are coming to be called clean-up filters.

127

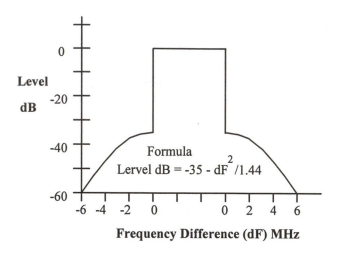

Figure 11.2: FCC Spurious Radiation Mask

Impedance Matching

One of the most difficult, and ultimately one of the most successful, tasks facing the developers of the analog transmission systems was the extent to which impedance matching had to be perfected in high-powered components such as feeders, antennas, and switching frames. This arose because mismatches gave rise to transmitted echoes that appeared on the screen as trailing ghosts. The echoes had to be sufficiently small to make the ghosts invisible to ordinary viewers. Thus antenna return losses of better than 30 dB were the order of the day.

With digital the situation is quite different. In Chapter 3 on radio frequency propagation we showed that all digital systems must be designed to be largely immune to the effects of the multipath propagation found in typical urban environments. Accordingly, the ATSC system includes an adaptive equalizer to cancel multipath echoes, and the COFDM systems employ a variety of techniques, including guard bands. Now, there is nothing to distinguish an echo due to an impedance mismatch from one due to multipath propagation, except that the former may be simpler because it is static in time and constant in level. It follows that digital systems have a very high immunity against mismatch echoes, and specifications for impedance matching can be considerably relaxed from those prevailing in the analog era.

One cannot take this too far, of course, because these echoes will absorb some of the systems' capacity to counter echoes arising in the external propagation path. However, it does follow that the specifications applying to transmission plant items designed for the analog service will be more than adequate for digital operation.

System Testing

The very essence of analog transmission systems is that they should faithfully reproduce all sorts of waveforms faithfully. Hence a large component of testing an analog system consists in sending standardized waveforms through it and measuring the difference, or distortion component, by which the reproduced version differs from the original. Thus, in analog TV we had test signals comprising square waves, luminance pulses, chrominance pulses, and the like.

We saw in Chapter 2 on the introduction to digital systems that the function of a digital transmission system is also to reproduce faithfully one particular signal, that comprising the symbol waveform for the system. Conceptually then, one might expect the digital system to be tested with a test waveform much as analog systems are. However, the only points at which distortion on this waveform is important is at the receiver sampling instances; at other times the distortion is quite incidental to the operation of the system. Now, as described in Chapter 2, any distortion affecting the signal at the sampling instances will cause a spreading of points on the constellation diagram as shown in Figure 11.3 for a QPSK signal.

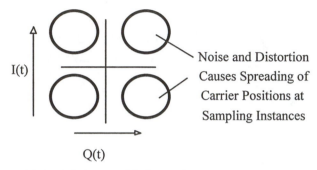

Figure 11.3: Noise and Distortion on Constellation Diagram

Figure 11.3 leads directly to a method of measuring the distortion a system may be producing. If we transmit a random data stream through the system, it is possible to determine the average distance by which all the observed constellation points differ from their correct positions. This is known as the error vector magnitude (EVM). The reason for the name will be obvious from the diagram in Figure 11.4 which illustrates the concept.

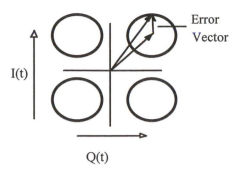

Figure 11.4: Concept of the Error Vector Magnitude

The error vector magnitude is likely to become the single most important measure of transmitter system performance. Specialized monitoring sets known as vector signal analyzers are becoming available and can continually monitor the transmitter while it is carrying program. Naturally, however, the general tests applicable to any radio system, such as the measurement of frequency stability and spurious radiation, will still be required.

We should make one point here about the use of vector signal analyzers with 8-VSB transmitters. An operator might expect to see a constellation diagram like that in Figure 11.5.a since, after all, there are only eight permitted signal levels. In fact, however, the instrument will show a diagram like that in Figure 11.5.(b). The former is applicable to straight linear amplitude modulation, but the conversion of the modulated signal to VSB introduces a component on the quadrature modulation axis described in Chapter 2. In this case the quadrature component does not carry any useful data, and its presence serves only to preserve the VSB nature of the signal. Readers familiar with the quadrature method of generating single sideband will recognize this effect,

131

and others may care to seek out a description of the method in standard texts.

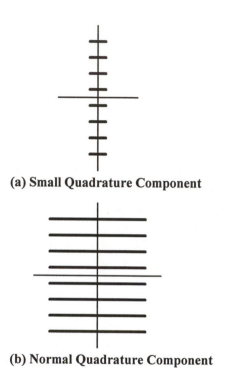

(a) Small Quadrature Component

(b) Normal Quadrature Component

Figure 11.5: Constellation Diagrams for 8-VSB

The quadrature component does not have the property of being zero at successive sampling instances. This is not important as the quadrature component does not carry data and intersymbol interference within it is of no consequence. However, it does mean that its amplitude at the sampling instances does vary randomly depending on the preceding data sequence. Thus the possible

132

constellation positions for each of the eight transmission levels become straight lines rather than dots and the diagram becomes that shown in Figure 11.5.(b).

This has necessarily been only a cursory introduction to EVM measurement. Vector signal analyzers promise to become powerful diagnostic tools, and a perusal of the manufacturers' literature is worthwhile.

Chapter 12

Transmission Systems in Practice

Throughout the book we have looked at a number of the characteristics of practical digital broadcasting systems to illustrate the principles that underlie digital transmission. This chapter brings together in a concise form the features of the principal systems now coming into service around the world.

It must be said, however, that there are a great many details of these systems that it would be impractical to cover in one chapter, or perhaps even in one book. These are available in the applicable standards where the interested reader will find that the description of even seemingly simple matters like the structure of the transport streams takes up a dauntingly large number of pages. The intention of this chapter, and for the whole book, is to give the reader an overview of the principles involved and an appreciation of how they have been put into practice.

In fact only a very few systems can really be regarded as world standards, and as we shall see, these share many features. Hence, we will commence the discussion by describing the typical configuration found in all of these systems.

Typical Configuration

Figure 12.1 shows the typical transmitter end configuration for a digital broadcasting system. Here we show encoding of video and audio in a multichannel configuration. Of course this could well be for only one channel or for audio only in the simplest applications.

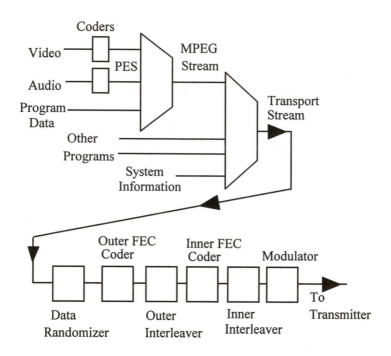

Figure 12.1: Typical Transmitter Configuration

The starting points for our universal transmission system are the coders that compress the incoming video and audio signals, giving the packetized elementary streams (PESs) that we discussed in Chapter 4. These streams are combined in a multiplexer with program associated data to give the MPEG transport stream. The transport streams from a number of sources may enter a second multiplexer together with any data streams and a new service channel we have not previously discussed, known as the system information (SI) channel.

The SI is essential in a multichannel system. It contains information on all the programs and data streams being carried. In particular, it tells the receiver what packet ID (PID) code is associated with each service. Thus, if a receiver is to be "tuned," to use the word loosely, to, say, program No. 2, it will decode only those transport packets carrying the appropriate PID and discard the rest. The SI carries other essential system information as well. For example, where the modulation parameters can be reconfigured, as in those systems employing COFDM, it can inform the receiver of impending changes to ensure that it alters its configuration at the correct instant. The SI is often transmitted with a more robust modulation method than the program signals themselves. For example, a smaller number of levels may be used in the digital symbol waveform while it is being transmitted. This is done to ensure that the SI is more immune than are the programs to interruption by adverse events such as noise bursts. Thus the receiver retains its correct settings and timings and the program recovery is rapid after such an event.

The multiplexer essentially combines all the input signals into the transport stream. The data stream from it feeds into the data randomizer, sometimes called the energy dispersal unit. Its purpose is to prevent the transmission of long strings of fixed or repetitive data such as, say, a repeating one-zero sequence or a long continuous run of ones or zeros. When these occur, they tend to make the signal out of the modulator sit on some fixed level or value, such that it resembles an unmodulated carrier. This can cause long-range interference to other services and is best prevented. The randomizer ensures that the modulation resembles a random signal and thus disperses the carrier energy over the transmission channel.

Then comes the outer coding for forward error correction (FEC). This is almost universally a Reed-Solomon cyclic redundancy block code as described in Chapter 5. After that we find the outer interleaver, which, as described in the same chapter, breaks up bursts of errors so that they become scattered single-bit errors in the data stream and are more amenable to correction by the FEC.

Following the outer interleaver, we come to the inner FEC coder, which is universally some type of convolutional or trellis coder. It may be worth mentioning that this inner coding is generally not used on cable distribution systems, as that environment is considered quite benign. Finally, we come to the outer interleaver, which, again as described in Chapter 5, tends to place the bits convolved together by the trellis coding far enough apart in time so that not all are affected by single noise burst.

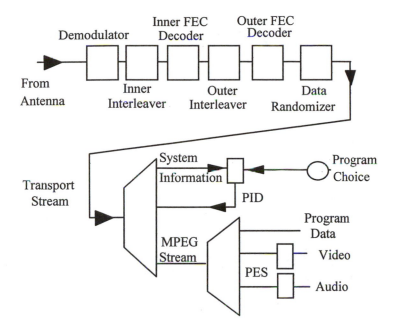

Figure 12.2: Typical Receiver Configuration

Figure 12.2 shows a typical receiver, which is in a sense the inverse of the transmitter diagram in Figure 12.1. Notice here that the program choice determines which PID is sought out by the transport demultiplexer and hence which transport stream packets are sent to the final MPEG demultiplexer.

It is in the modulation scheme that one finds the real differences among the standards that currently exist. The ATSC system developed in the United States uses eight level VSB transmission which is described in detail in Chapter 8. The DVB terrestrial system uses COFDM, as does the Eureka DAB system for sound broadcasting. COFDM is described in Chapter 9.

Satellite TV systems universally use QPSK as described in Chapter 10. The following paragraphs list the major characteristics of these systems in turn.

The ATSC 8-VSB System

The ATSC system closely reflects our universal digital system as described. In this case the video compression system is MPEG-2, and the audio compression is the Dalby Laboratories AC-3. We have already seen many details of the ATSC system as it was used in Chapter 8 to explain many of the principles underlying VSB transmission. Table 12.1 therefore lists only its major capabilities as a transmission system. The figures apply to the system operating in a 6 MHz channel as employed in North America.

Note that there is an extension to the standard for 16-level VSB modulation for application in cable systems. This achieves twice the bit rate at 38.6 Mb/s but has a carrier-to-noise (C/N) threshold of over 28 dB.

The term R-S(207,187) defining the outer FEC arrangement in Table 12.1 indicates that the FEC is Reed-Solomon in which the blocks comprise 207 bytes, of which 187 are message bytes and 20 are check bytes, as described in Chapter 5 on error correction.

Table 12.1: ATSC System Characteristics

Item	8-VSB	16-VSB	Units
Bandwidth	6.0	6.0	MHz
Symbol Rate	10.76	10.76	megasymbols/s
Outer FEC	R/S(207,187)	R/S(207,187)	
Inner FEC Rate	2/3	1	
Payload Data Rate	19.3	38.6	Mb/s
C/N Threshold	15.0	28.3	dB

It is interesting to see how the comparative bit rates in these two systems arise. In Chapter 8 we showed how the symbol rate of 10.76 megasymbols per second arises from a channel bandwidth of 6 MHz. The 8-VSB system uses 8-level modulation and therefore can transmit 3 bits per symbol. The 16-VSB system uses 16-level modulation giving four bits per symbol. However, the 16-level system does not employ an outer trellis coder whereas the 8-level system uses a rate two-thirds coder as explained in Chapter 8. Dividing the four-thirds factor for the extra levels by the two-thirds factor for FEC we see that the 16-level system should have twice the bitrate capacity of the 8-level system, as indeed Table 12.1 shows it does. However, the price is a loss of over 13 dB in noise immunity.

DVB Terrestrial System

This system uses COFDM modulation as described in Chapter 9. Both video and audio compression are MPEG. It is not

141

so easily described as the ATSC system, as the standard permits a great deal of variation. Most important, there are two choices for the number of carriers in the COFDM ensemble, giving what is known as the 2K or 8K modes. The number of carriers in the 2K mode is 1705, and in 8K mode it is 6817. Notice that this is nearly 2000 and nearly 8000, respectively, and the reason for the nomenclature 2K and 8K may be obvious.

Table 12.2 gives some critical values for what are regarded as fairly normal configurations.

Table 12.2: DVB Terrestrial System Characteristics

	8000 Carrier Mode	2000 Carrier Mode
No. of Carriers	6817	1705
Carrier Spacing	1.1 kHz	4.5 kHz
Symbol Duration	896 usec	224 usec

Typical Data Rates

	QPSK	16 QAM	64 QAM
FEC = 1/2	5.9	11.7	17.6
FEC = 2/3	7.8	15.6	23.4
FEC = 3/4	8.8	17.6	26.4

Mb/s with 1/16 Guard Interval

142

As Table 12.2 indicates, the modulation of the carriers in the COFDM ensemble may be changed between QPSK, 16 QAM and 64 QAM. The forward error correction may also be varied from rate $^1/_2$ to rate $^7/_8$, and finally, the guard interval may range over several values from $^1/_{32}$ to $^1/_4$ of the symbol interval. Hence, the number of possible permutations is enormous. This applies to European 8 MHz channels, and the standard permits direct scaling for other bandwidths.

DVB Satellite System

As outlined in Chapter 12, the DVB standard for satellite transmission specifies QPSK for the modulation system. It does, however, permit several values for the FEC rates. Table 12.3 shows the available data rates and required carrier-to-noise ratios for a typical transponder bandwidth of 33MHz. The bit rates can be scaled for different transponder bandwidths.

Table 12.3: DVB Satellite System Characteristics

FEC Rate	Payload (Mb/s)	C/N Reqd.
1/2	23.8	4.1 dB
2/3	31.7	5.8 dB
3/4	35.6	6.8 dB

For 33 MHz Transponder Bandwidth

Digital Techniques in Broadcasting Transmission

Eureka 147 DAB System

The Eureka 147 for digital audio broadcasting has essentially the same block diagram as that shown for the universal digital broadcasting system and lacks only the video encoder. Of course the bit rates are quite a lot lower than in the TV transmission systems and the required channel bandwidth is only 1.5 MHz. The modulation system is again COFDM. The standard describes four modes of operation, with the characteristics as shown in Table 12.4. The four modes arise because the relative immunities to multipath echoes, Doppler shift, and local oscillator instability depend on the symbol durations and carrier spacings in different ways. A broadcaster can choose the mode that best suits his circumstances.

Table 12.4: Eureka 147 DAB Characteristics

	Mode 1	Mode 2	Mode 3	Mode 4	
Max Transmitter Freq.	300	1500	3000	1500	MHz
No. of Carriers	1536	384	192	768	
Carrier Spacing	1	4	8	2	kHz
Symbol Duration	1246	312	156	623	usec
Guard Interval	246	62	31	123	usec

Satellite Sound Broadcasting Systems

There are effectively two systems recognized by the International Telecommunications Union for sound broadcasting from satellites. One, known as Digital System A, corresponds simply to transmitting the Eureka 147 ensemble as described above through a satellite. The other, known as Digital System B, is essentially the same as the Worldspace system described in Chapter 10. It is said that System A should be preferred in the frequency bands below 3GHz, and System B for frequencies above that. This, however, seems an oversimplification, and the choice will depend on many factors, including, in particular, the affordability of receivers to the target audience.

Chapter 13

The Mixed Analog and Digital Environment

Although digital transmission technology has developed apace in recent years, one can be certain that it will not replace the existing analog system overnight. The viewing public has a great deal of money invested in the population of existing receivers. Many are not going to be easily persuaded to change over to new and potentially far more expensive sets, even if they can be convinced that the quality of what they are hearing and seeing is going to be enhanced. It is doubtful that improved quality alone is a sufficient driving force for a rapid public acceptance of new technology. Many may remember how long it took for color TV sets to displace the population of black and white models after the introduction of color TV transmissions—and color offered a major advance in perceived quality. Hence, we can expect that it will be a number of years before the analog network can be closed down, and during that time the digital and analog transmissions must coexist. In fact, those governments that have announced formal timetables for the introduction of digital TV are allowing generally from six to eight years for the changeover. Almost always there is the caveat that this is to be reviewed, suggesting some lack of confidence that even that timetable can be achieved.

This chapter examines the factors that apply to a mixed digital and analog environment. It looks at how transmission channels are being made available to fit the new services in with the analog services currently operating. There are issues of matching the analog and digital coverage areas and of controlling mutual interference between the two. The chapter addresses the use of external plant carrying mixed analog and digital signals. It makes the assumption that the broadcasting organizations currently running analog services will want to move into the digital domain and use the existing infrastructure, such as antennas, towers, and feeders, to the maximum possible extent.

The treatment is largely confined to television broadcasting. As noted in Chapter 12 in the discussion of practical digital systems, there is as yet only one standardized system for sound broadcasting; the Eureka 147 DAB system. Generally speaking it is not used in the same frequency bands as the existing analog systems, and so the question of how the new and old systems might operate together and share facilities and frequencies does not really arise. It may in the future, when DAB enroaches into Band II, currently occupied by FM services. However, it is generally envisaged that DAB will develop in other bands until it has reached the degree of acceptance and penetration at which the FM services may be closed down. However, some countries are allocating vacant TV channels at the top of Band III for DAB services right from the outset. Generally speaking, however, the transmitter powers are so much lower than for TV and the transmitted bandwidth so much less that the sharing of any significant infrastructure will seldom be necessary.

Once the in band on channel (IBOC) or in band adjacent channel (IBAC) systems mentioned in Chapter 1 are developed, then we will have a very intensive sharing of frequencies and plant between digital and analog sound services. It is probably true to say that devising successful methods of sharing frequencies with the analog services has been the difficulty causing most delay in the realization of a fully accepted IBOC or IBAC scheme. For the moment, however, let us look at the situation with TV, where there is much to interest those faced with embracing the new technology.

Channel Allocation

Historically, a number of constraints have applied to the allocation of analog channels in the same or adjacent service area. These have arisen from the need to avoid mutual interference between the services, which are manifested as artifacts visible to the viewer. Thus, for example, no channel could be allocated if its frequency range included the local oscillator frequency of receivers tuned to another channel in the service area. In this case, the interference generated by those local oscillators could appear as herringbone patterns on the screens of the receivers affected. In particular, adjacent channels, either one up or one down from another channel in the area were very strictly avoided. Beats between the carriers of these channels, produced in receivers, fall into the video band and can produce very annoying patterns on the screen. In general, the forbidden allocations were known as "taboo" channels.

With digital, much of this has changed rather dramatically. Principally, the digital channels carry no strong single carriers with their propensity to produce interference and beat patterns. In

149

analog, of course, we have the vision and sound carriers and the color subcarriers that are always present at high levels. Rather, the spectrum in the digital channels is quite diffuse, with the energy spread very evenly throughout the band of frequencies occupied by the transmission channel. Hence, digital signals are much less likely than analog to cause visible interference. At the same time, as we shall see, they can be made quite resistant to interference from analog channels. When interference does occur to a digital channel, its effect is to lower the signal-to-noise immunity of the system, rather than to produce a visible artifact. Hence, even moderate levels of interference may not be visible at all to the viewer.

Taken together, these characteristics permit channel allocations in a mixed digital and analog environment that would never have been contemplated in a totally analog environment. In particular, they have made it possible to support both analog and digital services in that block of spectrum, the broadcasting bands, that previously supported only the analog channels. Or should we say, they have made it possible to establish the digital services without seeking an additional allocation of spectrum, which would have been exorbitantly expensive or more likely, not forthcoming.

Most of the old taboo rules are no longer applicable. One of the more important is that of allocating adjacent channels. To make the discussion concrete, let us take a service area in which say channels 30 and 33 were allocated. Neither channels 31 or 32 could then have been used for analog services, but either or both can now be used for digital services. The same, of course, applies to channels 29 and 34. In these circumstances the major constraint would probably be the question of what channels are operating in the adjacent service areas. This raises the matter of what protection ratios are required between digital and analog channels in these circumstances.

The protection ratios that authorities around the world are likely to adopt depend very much on local circumstances. Not only do the standards used for both the digital and analog systems differ, but so do the definitions of what constitutes harmful interference. Some authorities use the concept of "threshold of visibility" (TOV), whereas others adopt a certain error rate in the digital channel. Nevertheless, we can generalize to some extent and the following list is intended as a guide.

The following shows the desired-to-undesired (D/U) ratio in decibels between the wanted (desired) service and the interfering (unwanted) service. Where this is negative it indicates that the level of the interfering signal can actually be higher than that of the wanted signal:

Interference Digital to Digital
      Co-channel—D/U of 15 to 20 dB
      Adjacent Channel— D/U of –40 to –45 dB
Interference Analog to Digital
      Co-channel—D/U of 3 to 6 dB
      Adjacent Channel – D/U of –25 to –30 dB

Interference Digital to Analog
      Co-channel—D/U of 35 to 40 dB
      Upper Adjacent Channel—D/U of –12 to –15 dB
      Lower Adjacent Channel—D/U of –8 to –12 dB

These are generally limiting values at which artifacts become visible or at which a defined error rate occurs. System designers would generally allow some safety margin beyond these figures.

Some interesting observations can be made. For instance, we mentioned that the spectra of digital signals are relatively diffuse, and in this respect they reflect the properties of random noise. Hence, the D/U ratios for co-channel interference from digital into digital or analog channels roughly reflect the signal-to-noise ratios at which the wanted signals would begin to fail. Notice that D/U ratios for interference from digital into analog are more critical with digital in the upper adjacent channel than when it occupies the lower adjacent channel. This reflects the fact that in all analog systems the sound carriers are situated very close to the upper channel edge in frequency and are accordingly somewhat sensitive to interference from the abutting upper channel. This effect will be critical when we come to discuss adjacent channel combining.

The digital systems do have the means to lessen the effects of analog interference. Thus the ATSC standard for 8-VSB describes a NTSC filter that may be switched in or out. This is a fairly simple comb filter with notches at the frequencies of the NTSC vision, sound, and color subcarriers. It does produce bursts of interference when fast transients occur in the interfering NTSC channel. These have a duration of about 12 DTV symbol periods and are among the reasons why the trellis FEC is spread over 12 symbols. For systems using COFDM it is possible to leave out or simply not modulate those carriers having frequencies close to the interfering carriers, and instruct the receiver to ignore anything on those frequencies. The D/U ratios quoted reflect the use of these techniques.

Options for Carrying Digital

Given the extent to which the properties outlined make digital channels available, it is highly likely that existing broadcasters will be allocated channels in the same band as their existing analog services, at least in the UHF bands and in Band III. This means that it may well be possible to share the existing infrastructure of antennas, feeders, and towers between the digital and analog services. A broadcaster in this position has essentially three options:

1. to combine both services into the existing or a new antenna;
2. to split the existing antenna into analog and digital halves;
3. to add a digital antenna to the tower.

There is, of course, the fourth option of completely replacing the existing infrastructure or of building a separate tower and antenna system. If one is able to afford this and obtain the necessary environmental clearances, then it is certainly the simplest technically, but the least interesting. Our purpose in this chapter, however, is to look into the factors that govern the more technically challenging solutions that most broadcasters will have to implement. Let us take up first the matter of channel combining and more particularly adjacent channel combining.

Adjacent Channel Combining

This section is headed adjacent channel combining rather than simply channel combining because the interesting problems occur in the former rather than in the latter. If one is able to successfully implement the combining of adjacent channels, then the combining of nonadjacent channels will seem simple by comparison.

Many readers of this book will not be familiar with the concept of channel combining as in some countries it has rarely found any application in the analog era. However, it offers some very attractive solutions for carrying dual digital and analog services during the time up to when the analog services cease. For this reason, Appendix 2 gives a brief outline of the theory of channel combiners and how they are realized in practice. There the reader will find that combiners possess wide-band and narrow-band ports. Typically, the narrow-band port couples one single channel through to the output, whereas the wide-band input provides a transparent connection through to the output for many channels.

The manner in which combining modules are connected to form a combining chain is shown in Figure 13.1. Here, to be concrete, we have shown channels 30, 31, and 32 being combined, but the numbers are arbitrary.

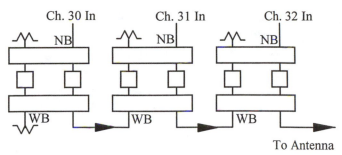

Ch. 30 In   Ch. 31 In   Ch. 32 In

To Antenna

Terminating Loads

NB = Narrow Band Port

WB = Wide Band Port

Figure 13.1: A Channel Combiner Chain

The important point about channel combiners is that they impose a filter characteristic on the channels that pass through both the narrow-band and broad-band ports.

Figure 13.2 shows the characteristics, in terms of amplitude and group delay, that are to be expected in the narrow-band and broad-band ports. The narrow-band port's attenuation and group delay characteristics accurately reflect those of the filters used in the combiner module, as the discussion of combiner operation in Appendix 2 shows. The wide-band port's response is effectively the inverse of this. There is a small transitional frequency range at each edge of the narrow-band port's passband over which there is some attenuation through both ports. We show in Appendix 2 that there can be no frequency at which the attenuation through both ports is zero, so that these transition regions must exist in all combiners. There is a frequency at which the attenuation through both ports is equal, and it would be 3 dB for both in an ideal combiner made of lossless components. This is often referred as

155

the 3 dB point of the combiner. As we move away from the 3 dB point in the transition region, the attenuation through one port falls and the attenuation through the other rises.

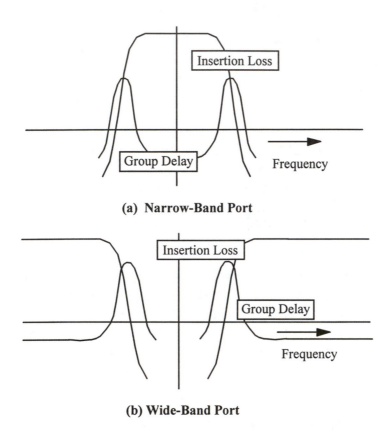

(a) **Narrow-Band Port**

(b) **Wide-Band Port**

Figure 13.2: Filter Characteristics of a Combiner Module

If the combiner is to be used for nonadjacent channels the transition regions cause few problems. They can be placed at frequencies not occupied by the channels being combined.

156

However, with adjacent channel combining, one transition region at least must occur over the frequency at which the channels are contiguous. Hence, both channels will be affected to some extent by attenuation and group delay.

There are two situations to consider here. The first, where the digital signal occupies the channel lower in frequency and the analog channel the upper, causes no particular problems. The second, however, with digital in the upper channel and analog in the lower, needs particular care in implementation.

One of the difficulties that arises is the attenuation of the sound carrier of the analog channel. We mentioned that in all the common analog transmission standards the sound carrier is very close to the channel edge. In fact, where stereo sound is concerned, one carrier may be only a few kilohertz from that edge. Now, if we place the combiner 3 dB attenuation point at the common channel edge, it is evident that there must be some significant attenuation of that sound carrier, since it lies so close to the point at which the port attenuation is 3 dB. The actual attenuation will depend on how steep the attenuation characteristic of the filters in the combiner can be made. Because of the limits imposed by power handling requirements and finite Q values in resonators, eight pole filters are about the steepest that can be used in practice. Hence, one finds that the attenuation of the sound carrier is nearly always 1.5 to 2 dB.

In practice, this means that the power output for the transmitter's sound carrier must be increased. On klystron transmitters this is rarely a problem, but on some solid-state transmitters it may actually be necessary to install additional aural amplifiers.

Of course, by the same argument there must be a similar attenuation of that part of the digital channel that also falls into the combiner transition region. The effects are difficult to analyze, but have been studied by computer simulation. This shows that the amplitude attenuation by itself has a reasonably negligible effect of the digital system performance. The real culprit is the group delay distortion.

In Figure 13.3 we reproduce the spectral characteristics of the ATSC 8-VSB signal. Notice that the carrier pilot falls only 310 kHz above the lower edge of the channel, which puts it well into the region of high group delay for all practical filters. Recall that high group delay indicates phase changing rapidly with frequency. Then, since the system uses the pilot as its phase reference it is not surprising this group delay has a very detrimental effect on system operation.

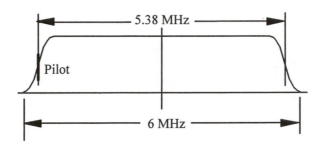

Figure 13.3: 8-VSB Channel Characteristic

Computer simulations have shown that if the group delay can be corrected the resulting system performance is quite satisfactory. The most practical method is to place an equalizer in front of the transmitter at baseband. It has been suggested that this could be an adaptive equalizer adopted from receiver practice. In that case, the training signal for the equalizer might be taken at RF from some point in the antenna feeders beyond the overall transmitter output. This realizes a closed loop correction system and in effect, continuously and dynamically corrects all the transmitter characteristics.

When the digital system is one of the COFDM variants, the effect of the group delay is probably negligible, as it will be small compared to the symbol duration on the outer carriers. However, it would be wise to check that this is so in unusual or critical situations.

Antenna Systems

We have previously made the point in this book that there is nothing special required of antenna systems for transmitting digital signals. The digital signals themselves are designed to be particularly robust against the principle problem of ghosting, which is the characteristic of antenna systems of most concern in analog transmission. Hence, it is possible that many existing antennas can be pressed into carrying both digital and analog services during the transition period. One needs to be cautious, however, because the combined bandwidth may be far greater than the original design requirement. Also, where adjacent channels are concerned, there is a requirement that the signal levels in the service area retain their correct ratios. This is to avoid unacceptable interference into the lower-level channel if the relative level of one of the adjacent channels should become much higher than the other.

The two common types of high-powered transmitting antennas are slot arrays and panel arrays. Slot arrays are known to have very narrow impedance bandwidths, inasmuch as an acceptable impedance match to the feeder cables can be achieved over only one or two TV channels. Hence they would be usable only where the digital and analog allocations specify adjacent channels, or sometimes, separation by one vacant channel.

This limitation on slot arrays arises largely because of the nature of the feed system, which is restricted by the small mechanical cross section of the antenna to having the radiating elements connected in sequence or series. By contrast, panel arrays with their greater mechanical cross sections accommodating complex sets of cables can exploit parallel feeding and associated impedance compensation techniques. It seems not often realized that the two different feed techniques also have a major influence on the susceptibility of the radiation pattern to vary with frequency, as we shall now show.

There are three methods of arranging the power distribution to the large number of levels of radiating elements found in high-gain arrays. These are shown in Figure 13.4 and comprise:

1. End feeding, whereby the feeder enters the bottom of the antenna and power to the radiators is tapped off the cable as it ascends the antenna column.

2. Center feeding, where the main cable enters the array at its mid-height and splits into one descending and one ascending feeder within the column from which power is tapped off to the radiators in sequence, as before.

3. Parallel feeding, whereby the feed cables split regularly and form a fanlike arrangement spreading over the height of the column. It is readily seen that all the radiating elements are thus fed in parallel, or more importantly, the electrical path lengths to the radiators from the common feeder are all essentially equal.

In end and center feeding, the taps off the feed lines are arranged to be at intervals of one wavelength at the design frequency so that all the radiators are correctly phased at this frequency. As mentioned, end and center feeding are necessarily used in slot arrays to minimize mechanical size, whereas parallel feeding is almost always exploited in broadband panel arrays.

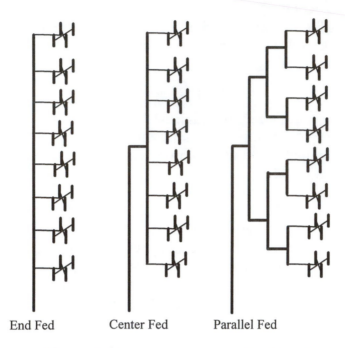

End Fed   Center Fed   Parallel Fed

Figure 13.4: Antenna Array Feed Methods

Mechanically, the arrangement of the feed systems in panel arrays is relatively straightforward, as the space allows the use of semi-flexible coaxial feeders formed into cable harnesses. In slot arrays, the slotted external cylinder commonly forms the outer of a transmission line, with the slots being inductively coupled to an internal tubular conductor located coaxially within the outer cylinder and forming the inner of the line. For center feeding, there may be a third conductor inside the lower half of the inner conductor, forming a triaxial line connecting to the common feed point at the center of the antenna.

Vertical Radiation Patterns

In looking at Figure 13.4 it is intuitively obvious that the relative phases to the radiating elements will vary with frequency more in the end-fed arrangement than in the center fed, and this in turn will vary more than in the parallel feed system. In the end-fed array the shortest and longest cables from the common feed point differ in length by an amount equal to the length of the antenna column. In high-gain antennas this may be equal to some 25 to 30 wavelengths. Obviously, the phase difference over this length will change markedly with frequency and cause a frequency-dependent variation of the vertical pattern.

By contrast, in the parallel array, the lengths of all the cables from the common feed point are all essentially equal and no relative phase changes are to be expected as the frequency is varied. The center-fed array has characteristics somewhere between the two.

Figure 13.5 shows the extent to which these phase changes can cause the vertical pattern to vary with frequency. Figure 13.5.a is for the end-fed case, 13.5.b for the center-fed case, and 13.5.c for the parallel-fed array. In each case the design frequency of the antenna is 600 MHz and the plots are taken at this frequency and +/- 6 MHz. The antenna mid band gain is 15 dB; typical of many applications.

163

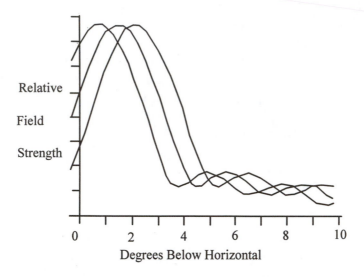

Figure 13.5.a: VP Variation - End-Fed Array

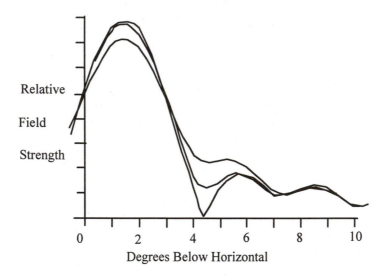

Figure 13.5.b: VP Variation - Center-Fed Array

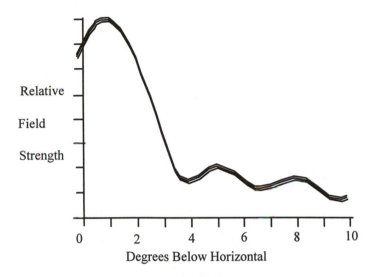

Figure 13.5.c: VP Variation - Parallel-Fed Array

Appendix 1

The Fourier Transform

In thinking about digital transmission it is very useful to be able to change one's point of view back and forth between the time and the frequency domains. In other words one switches between thinking about waveforms and thinking about spectra. A drawing of a waveform would be a plot of the voltage along a time axis. A drawing of its spectrum would show a certain level plotted along a frequency axis. One can think of this level as representing the energy contained in a very thin slice of frequencies surrounding any given point on the frequency axis. The mathematical process by which one can convert between waveforms and spectra is known as the Fourier transform.

The name derives from that of its inventor Jean Baptiste Fourier, who in 1822 realized that a periodic waveform could be represented by the sum of a number of sine waves with frequencies related to the periodicity of the waveform. The expression giving the amplitude of each sine wave became known as the Fourier series. The Fourier transform is an extension of this. The Fourier series gives a voltage representing the amplitude of discrete sine waves at each frequency involved. The transform on the other hand returns a voltage that, if squared, would give the energy flow around any given frequency. Hence, there is a one-to-one correspondence between the spectrum and the Fourier transform.

The Fourier transforms of many common waveforms have been derived and are available in the literature. Figure A1.1 shows the transform of a rectangular pulse. The transform has the mathematical form of sin(x)/x, referred to as the sinc pulse or more often as the "sine x on x" pulse. The transform of a series of sharp impulses is itself a series of impulses, as shown in Figure A1.2. We have used this relationship in the text to explain the Sampling Theorem, and why high quality sound must be sampled at at least 40kHz.

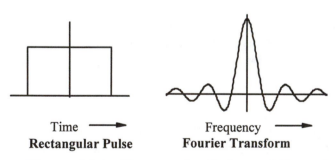

Time $\longrightarrow$ Frequency $\longrightarrow$

Rectangular Pulse   **Fourier Transform**

Figure A1.1: Rectangular Pulse and Transform

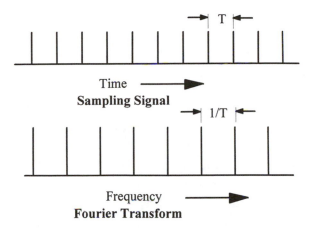

Time ──────▶
Sampling Signal

Frequency ──────▶
Fourier Transform

Figure A1.2: The Sampling Signal and Transform

An interesting thing about Fourier transforms is that they are reciprocal. Thus in Figure A1.3 we show a sinc pulse in time and its spectrum, which turns out to be rectangular in form. Notice that this is just the reverse of a rectangular pulse and its spectrum in Figure A1.1.

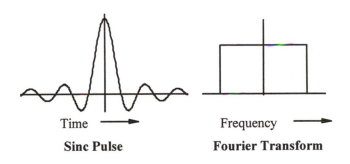

Time ──────▶   Frequency ──────▶

Sinc Pulse   **Fourier Transform**

Figure A1.3: Sinc Pulse and Transform

Another important point is that a single isolated voltage impulse of essentially zero width in time has a flat spectrum ideally stretching from minus infinity to plus infinity in the frequency domain. Now if we were to pass this pulse through a filter with a rectangular frequency response, the filter would let through only that part of the spectrum that falls within its pass band. In other words the output spectrum would be rectangular, as shown in Figure A1.3, and hence the output pulse must be a sinc pulse, just as shown in that figure. This, as discussed in Chapter 2, is the very basis, the *sine qua non*, of digital transmission.

Principles of Channel Combiners

As digital TV commences many existing operators of single analog services may find themselves faced for the first time with channel combining. Even if it is unfamiliar, channel combining may well be the system of choice for running the old analog and the new digital services in parallel during the transition period leading up to a fully digital environment. Where combining is possible it will almost certainly be cheaper than installing a separate digital antenna with all the implications for tower upgrading. Hence, this appendix is meant to provide a short review of modern combiner principles for those who may not have previously encountered the technique.

Modern channel combiners are based on a circuit known as the constant impedance ring, sometimes called the Lorenz ring after its inventor. As we will show, it has the major advantage of presenting a constant impedance to the transmitter, both inside and outside the transmission channel, and as such minimizes the production of spurious radiation in the transmitter amplifiers. We mentioned in Chapter 11, Engineering the Transmission Channel, that controlling spurious radiation is likely to become a major issue, especially during the transition period when analog and digital services must coexist in limited frequency bands. Accordingly, not only must the generation of spurious radiation be controlled, but also it is highly likely that most high-powered transmitters will have to be fitted with output filters to achieve the mandatory specification. These filters, called clean up filters, are also constructed in the form of constant impedance rings. Hence, we can expect these circuits to find wide application in the immediate future.

The 3 dB Directional Coupler

The major circuit component involved in forming the constant impedance ring is the 3 dB directional coupler. Hence, let us begin by reviewing signal flow through these simple devices. Figure 2.1 shows the necessary flow diagrams.

Figure A2.1 shows the relationships between the voltages of the signals at each port when one port is driven and the other three are terminated in matched loads. In this diagram the use of the letter "j" implies a phase shift of 90 degrees, in accordance with common practice in representing voltages in circuits.

Take diagram (a) in that figure, for example. Here we see that when a voltage V excites port 1, a voltage $V/\sqrt{2}$ emerges from port 2 and one of $jV/\sqrt{2}$ from port 3. Nothing emerges from port 4, and we say that ports 1 and 4 are isolated. Similarly, diagrams (b) through (d) in that figure show the signals at each port when ports 2, 3, and 4 are excited.

172

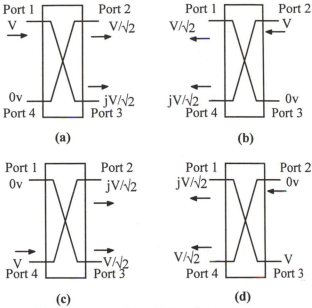

Figure A2.1: Signal Flow in Couplers

Now consider a coupler configured as in Figure A2.2. Here, port 1 is excited by a voltage V, ports 2 and 3 are terminated in open circuits, and port 4 is terminated in a matched load. By virtue of the drive at port 1, signals with voltages $V/\sqrt{2}$ and $jV/\sqrt{2}$ emerge from ports 3 and 4. These are immediately reflected by the open circuits and reenter the coupler. Now consider the signal reentering port 2. Because the coupler action is the same as illustrated in Figure A2.1.b and described previously, it must generate a signal of voltage $V/2$ at port 1 and $jV/2$ at port 4. Similarly, the signal with voltage $jV/\sqrt{2}$ reflected at port 3 must produce a voltage $jV/2$ at port 4 and $-V/2$ at port 1. This minus sign arises because this particular component has been subjected to two 90-degree phase shifts in its two passages through the coupler. A combination of bold and normal lettering is used on the diagram to assist the reader to trace the component signal flows. Notice now that these component signals add in phase at port 4, but cancel at port 1.

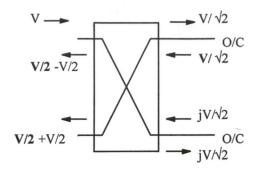

Figure A2.2: Reflections in a Coupler

The ultimate result of the coupler action in this case is to direct all the input power out of port 4. Also, since the net voltage emerging from port 1 is zero, there is no reflected power, and the coupler is matched at this input port.

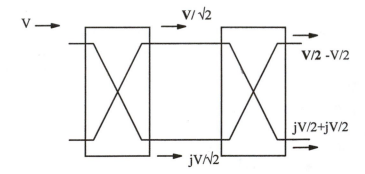

Figure A2.3: Cascaded Couplers

Consider two of the couplers connected together as shown in Figure A2.3. Again we use bold and normal lettering to help with tracing the signal flow. Essentially the signal with voltage $V/\sqrt{2}$ entering port 1 of the right-hand coupler produces an output $V/2$ at port 2, and $jV/2$ at port 4 of that coupler. The signal $jV/\sqrt{2}$ entering port 4 gives rise to voltages $-V/2$ at port 2 and $jV/2$ at port 3. The signals cancel at port 2 and add in phase at port 3. Hence the net result is that all the input power emerges at port 3. All the other ports are isolated from the input signal.

An important question is whether the two lines between the two couplers are matched, since, as we shall see, this could affect the characteristics of the filters to be placed in these lines. To address the question, look at the signal flows into the right-hand coupler. The two inputs are into ports 1 and 4. However, as discussed, these two ports are isolated from each other in normal coupler operation. Hence, no signals emerge from these ports, and none of the power entering is reflected. Hence, by definition, these two ports are matched. We are now in a position to look at how a modern constant impedance combiner operates.

The Constant Impedance Combiner

Figure A2.4 shows the circuit for the constant impedance combiner. It will be obvious why the term *ring* is used to describe this structure. The two filters shown in the ring are identical narrow-band bandpass devices. They define the passband for the port labeled as the narrow-band port on the figure.

175

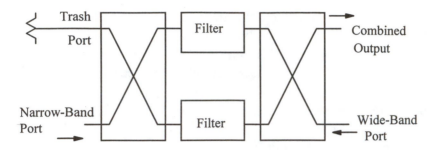

Figure A2.4: The Constant Impedance Combiner

Consider a signal within the passband entering this port. As the filters are effectively transparent to this signal, we have exactly the situation shown in Figure 2.3, and the signal passes through to the output port as explained for that figure. Now consider a signal entering the wide-band port. If it is outside the filters' passband, it will be reflected from the filters, just as it would be by the open or short circuits discussed in relation to Figure 2.2. Accordingly, as explained for that diagram, all of the input power will now emerge from the output port also. Hence, here we have the essentials of the combiner. The signals into the narrow-band port are within the filter bandwidth, the signals into the wide-band port lie outside that bandwidth, and the combiner acts to direct both to a common load.

A few additional points illustrate some of the important properties of these devices. For example, let us suppose the filters exhibit some small mismatch in the passband, as will inevitably be the case with practical devices. This would mean that some signal is reflected back into the output ports of the coupler shown on the left-hand side in Figure A2.4. By tracing the signal flow, as we did for Figure 2.2, it is easy to show that these reflected signals cancel at the input port, but add in phase at the port labeled the trash port in Figure A2.4. Hence, the input remains matched, and this reflected power is directed into the balancing load where it is absorbed. Generally speaking, for a well-set-up combiner, this power will never exceed a few tens of watts in the highest power applications. Similarly, if some signal from the wide-band port happens to leak past the filters, it can easily be shown to finish up in the balancing load also.

We need to qualify this simple picture a little. If the reader cares to trace the signal flows carefully, he or she will find that it is the sum of the two leakage signals that ends up in the balancing load. The difference is directed into the transmitter connected to the narrow-band port, and contributes a loss of isolation between transmitters. Similarly, if the in-band reflections from the two filters discussed above are not identical, the difference also is directed back toward the transmitter, causing some mismatch. The primary requirement in designing and setting up these combiners is to match the characteristic of the two filters to the best extent possible.

We have discussed the operation of the Lorenz ring from the point of view of combining, but as mentioned, it also finds much application in transmitter output filters. It will be evident from the discussion so far that the narrow-band port is matched over a broad frequency range, much wider than the operating channel bandwidth. This occurs because all the energy reflected from the internal filters finishes up in the balancing load, whether or not it falls within the passband of the filters. Thus, the transmitter sees a broadband-matched resistive load, as distinct from the highly reactive load it would see from an ordinary filter. The design and adjustment of the power amplifier are much simpler than they might otherwise be.

The Impossible Combiner

In Chapter 13, on operating combined analog and digital services into one antenna, we stated that some attenuation in each channel is unavoidable. This seems to disturb many people, and the question frequently arises as to whether some clever solution may be found whereby one could realize a perfect combiner with no attenuation in either channel. Some people have spent many hours searching for a realization of this perfect circuit. We shall now show that this ideal is impossible, even in principle.

To show that this is so, we will adopt a form of argument often exploited by mathematicians; that is, we will assume that a solution exists and then show that it leads to ridiculous consequences. Hence, let us suppose that the ideal combiner can be made, and that it will show no attenuation from either input port through to the output. We will call the input ports input 1 and input 2 in this case, since it would not make sense to keep labeling them as narrow band and broadband.

Consider the setup shown in Figure A2.5. Here we have an oscillator with a split output feeding signals with voltage V into each input port. As the combiner ports are matched, the power into each is V^2/R, where R is the characteristic impedance, and hence the total power entering the combiner is $2V^2/R$.

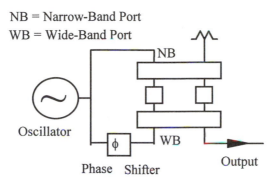

NB = Narrow-Band Port
WB = Wide-Band Port

Oscillator

Phase   Shifter

Output

Figure A2.5: Ideal Combiner Drive Conditions

The voltage at the output due to that at input 1 must be V, since by definition our ideal combiner must have no attenuation. Similarly, the voltage at the output due to that at input 2 must also be V. However, these two signals are at the same frequency, since they are generated in the one oscillator. By adjusting the phase shifter in Figure A2.5, we could bring them into phase, whereby the combined amplitude would be 2V. The power coming out of the combiner would be $4V^2/R$, against the $2V^2/R$ going in. This is clearly impossible, and the conclusion must be that our ideal combiner cannot be realized.

One can refine this argument by assuming that the voltages at the two input ports are unequal in amplitude. The result always shows that if at some frequency the losses from one port are small, then those from the other must be large. Also, if the losses are equal at some frequency, then they cannot be less than 3 dB. This is the basis for the statement made in the book that, for adjacent channel combining, the attenuation at the common channel boundary will be at least 3 dB.

Glossary

ADTV   Advanced Digital TV—an early name for the
                  United States Digital System

ATSC   Advanced Television Systems Committee—
                  United States Committee developing standards
                  for DTV

b/s   Bits per second

CCIR   Initials of the French name for the organization
                  preceding the ITU-R

COFDM   Coded Orthogonal Frequency Division Multiplex

CTV   Cabled TV Distribution System

DAB   Digital Audio Broadcasting

DTTB   Digital Terrestrial Television Broadcasting

DTV	Digital TV
DVB	Digital Video Broadcasting—a consortium, originally European, developing standards for DTV broadcasting
ERP	Equivalent Radiated Power
ETS	European Telecommunications Standard
ETSI	European Telecommunication Standards Institute
Eureka 147	European project to develop DAB
FCC	Federal Communications Commission
Grand Alliance	Consortium of United States manufacturers developing ADTV
ITU	International Telecommunications Union
ITU-R	International Telecommunications Union - Radio
kb/s	Kilobits per second
Mb/s	Megabits per second
Megasymbol	One million symbols
NTSC	World standard for analog TV originally developed in the United States

OFDM   Orthogonal Frequency Division Multiplex

PAL   World standard for analog TV originally

          developed in Europe

Symbol   One isolated digital signaling waveform

INDEX

186